D0141630

Alternative Energy

Alternative Energy

Volume 2

Neil Schlager and Jayne Weisblatt, editors

U•X•L

An imprint of Thomson Gale, a part of The Thomson Corporation

Detroit • New York • San Francisco • San Diego • New Haven, Conn. • Waterville, Maine • London • Munich

Alternative Energy

Neil Schlager and Jayne Weisblatt, Editors

Project Editor
Madeline S. Harris

Editorial
Luann Brennan, Marc Faeber, Kristine Krapp, Elizabeth Manar, Kim McGrath, Paul Lewon, Rebecca Parks, Heather Price, Lemma Shomali

Indexing Services
Factiva, a Dow Jones & Reuters Company

Rights and Acquisitions
Margaret Abendroth, Timothy Sisler

Imaging and Multimedia
Randy Bassett, Lezlie Light, Michael Logusz, Christine O'Bryan, Denay Wilding

Product Design
Jennifer Wahi

Composition
Evi Seoud, Mary Beth Trimper

Manufacturing
Wendy Blurton, Dorothy Maki

For more information, contact
Thomson Gale
27500 Drake Rd.
Farmington Hills, MI 48331-3535
Or you can visit our Internet site at
http://www.gale.com

Cover photograph for Volume 1 Oil drilling platforms off the coast of Texas © Jay Dickman/Corbis.

LIBRARY OF CONGRESS CATALOGING-IN-PUBLICATION DATA

Alternative energy / Neil Schlager and Jayne Weisblatt, editors.
p. cm.
Includes bibliographical references and index.
ISBN 0-7876-9440-1 (set hardcover : alk. paper) –
ISBN 0-7876-9439-8 (vol 1 : alk. paper) –
ISBN 0-7876-9441-X (vol 2 : alk. paper) –
ISBN 0-7876-9442-8 (vol 3 : alk. paper)
1. Renewable energy sources. I. Schlager, Neil, 1966- II. Weisblatt, Jayne.

TJ808.A475 2006
333.79'4–dc22 2006003763

This title is also available as an e-book
ISBN 1-4414-0507-3
Contact your Thomson Gale sales representative for ordering information.

Printed in China
10 9 8 7 6 5 4 3 2 1

Contents

Introduction

Alternative Energy offers readers comprehensive and easy-to-use information on the development of alternative energy sources. Although the set focuses on new or emerging energy sources, such as geothermal power and solar energy, it also discusses existing energy sources such as those that rely on fossil fuels. Each volume begins with a general overview that presents the complex issues surrounding existing and potential energy sources. These include the increasing need for energy, the world's current dependence on nonrenewable sources of energy, the impact on the environment of current energy sources, and implications for the future. The overview will help readers place the new and alternative energy sources in perspective.

Each of the first eight chapters in the set covers a different energy source. These chapters each begin with an overview that defines the source, discusses its history and the scientists who developed it, and outlines the applications and technologies for using the source. Following the chapter overview, readers will find information about specific technologies in use and potential uses as well. Two additional chapters explore the need for conservation and the move toward more energy-efficient tools, building materials, and vehicles and the more theoretical (and even imaginary) energy sources that might become reality in the future.

ADDITIONAL FEATURES

Each volume of *Alternative Energy* includes the overview, a glossary called "Words to Know," a list of sources for more information, and an index. The set has 100 photos, charts, and illustrations to enliven the text, and sidebars provide additional facts and related information.

ACKNOWLEDGEMENTS

U•X•L would like to thank several individuals for their assistance with this set. At Schlager Group, Jayne Weisblatt and Neil Schlager oversaw the writing and editing of the set. Michael J. O'Neal, Amy Hackney Blackwell, and A. Petruso wrote the text for the volumes.

In addition, U•X•L editors would like to thank Dr. Peter Brimblecombe for his expert review of these volumes. Dr. Brimblecombe teaches courses on air pollution at the School of Environmental Sciences, University of East Anglia, United Kingdom. The editors also express their thanks for last minute contributions, review, and revisions to the final chapter on alternative and potential energy resources to Rory Clarke (physicist, CERN), Lee Wilmoth Lerner (electrical engineer and intern, NASA and the Fusion Research Laboratory at Auburn University), Larry Gilman (electrical engineer), and K. Lee Lerner (physicist and managing director, Lerner & Lerner, LLC).

COMMENTS AND SUGGESTIONS

We welcome your comments on *Alternative Energy* and suggestions for future editions of this work. Please write: Editors, *Alternative Energy*, U•X•L, 27500 Drake Rd., Farmington Hills, Michigan 48331-3535; call toll free: 1-800-877-4253; fax: 248-699-8097; or send e-mail via www.gale.com.

Words to Know

A

acid rain: Rain with a high concentration of sulfuric acid, which can damage cars, buildings, plants, and water supplies where it falls.

adobe: Bricks that are made from clay or earth, water, and straw, and dried in the sun.

alkane: A kind of hydrocarbon in which the molecules have the maximum possible number of hydrogen atoms and no double bonds.

anaerobic: Without air; in the absence of air or oxygen.

anemometer: A device used to measure wind speed.

anthracite: A hard, black coal that burns with little smoke.

aquaculture: The formal cultivation of fish or other aquatic life forms.

atomic number: The number of protons in the nucleus of an atom.

atomic weight: The combined number of an atom's protons and neutrons.

attenuator: A device that reduces the strength of an energy wave, such as sunlight.

B

balneology: The science of bathing in hot water.

barrel: A common unit of measurement of crude oil, equivalent to 42 U.S. gallons; barrels of oil per day, or BOPD, is a standard measurement of how much crude oil a well produces.

biodiesel: Diesel fuel made from vegetable oil.

bioenergy: Energy produced through the combustion of organic materials that are constantly being created, such as plants.

biofuel: A fuel made from organic materials that are constantly being created.

biomass: Organic materials that are constantly being created, such as plants.

bitumen: A black, viscous (oily) hydrocarbon substance left over from petroleum refining, often used to pave roads.

bituminous coal: Mid-grade coal that burns with a relatively high flame and smoke.

brine: Water that is very salty, such as the water found in the ocean.

British thermal unit (Btu or BTU): A measure of heat energy, equivalent to the amount of energy it takes to raise the temperature of one pound of water by one degree Fahrenheit.

butyl rubber: A synthetic rubber that does not easily tear. It is often used in hoses and inner tubes.

C

carbon sequestration: Storing the carbon emissions produced by coal-burning power plants so that pollutants are not released in the atmosphere.

catalyst: A substance that speeds up a chemical reaction or allows it to occur under different conditions than otherwise possible.

cauldron: A large metal pot.

CFC (chlorofluorocarbon): A chemical compound used as a refrigerant and propellant before being banned for fear it was destroying the ozone layer.

Clean Air Act: A U.S. law intended to reduce and control air pollution by setting emissions limits for utilities.

climate-responsive building: A building, or the process of constructing a building, using materials and techniques that take advantage of natural conditions to heat, cool, and light the building.

coal: A solid hydrocarbon found in the ground and formed from plant matter compressed for millions of years.

coke: A solid organic fuel made by burning off the volatile components of coal in the absence of air.

cold fusion: Nuclear fusion that occurs without high heat; also referred to as low energy nuclear reactions.

combustion: Burning.

compact fluorescent bulb: A lightbulb that saves energy as conventional fluorescent bulbs do, but that can be used in fixtures that normally take incandescent lightbulbs.

compressed: To make more dense so that a substance takes up less space.

conductive: A material that can transmit electrical energy.

convection: The circulation movement of a substance resulting from areas of different temperatures and/or densities.

core: The center of the Earth.

coriolis force: The movement of air currents to the right or left caused by Earth's rotation.

corrugated steel: Steel pieces that have parallel ridges and troughs.

critical mass: An amount of fissile material needed to produce an ongoing nuclear chain reaction.

criticality: The point at which a nuclear fission reaction is in controlled balance.

crude oil: The unrefined petroleum removed from an oil well.

crust: The outermost layer of the Earth.

curie: A unit of measurement that measures an amount of radiation.

current: The flow of electricity.

D

decay: The breakdown of a radioactive substance over time as its atoms spontaneously give off neutrons.

deciduous trees: Trees that shed their leaves in the fall and grow them in the spring. Such trees include maples and oaks.

decommission: To take a nuclear power plant out of operation.

dependent: To be reliant on something.

distillation: A process of separating or purifying a liquid by boiling the substance and then condensing the product.

distiller's grain: Grain left over from the process of distilling ethanol, which can be used as inexpensive high-protein animal feed.

drag: The slowing force of the wind as it strikes an object.

drag coefficient: A measurement of the drag produced when an object such as a car pushes its way through the air.

E

E85: A blend of 15 percent ethanol and 85 percent gasoline.

efficient: To get a task done without much waste.

electrolysis: A method of producing chemical energy by passing an electric current through a type of liquid.

electromagnetism: Magnetism developed by a current of electricity.

electron: A negatively charged particle that revolves around the nucleus in an atom.

embargo: Preventing the trade of a certain type of commodity.

emission: The release of substances into the atmosphere. These substances can be gases or particles.

emulsion: A liquid that contains many small droplets of a substance that cannot dissolve in the liquid, such as oil and water shaken together.

enrichment: The process of increasing the purity of a radioactive element such as uranium to make it suitable as nuclear fuel.

ethanol: An alcohol made from plant materials such as corn or sugar cane that can be used as fuel.

experimentation: Scientific tests, sometimes of a new idea.

F

feasible: To be possible; able to be accomplished or brought about.

feedstock: A substance used as a raw material in the creation of another substance.

field: An area that contains many underground reservoirs of petroleum or natural gas.

fissile: Term used to describe any radioactive material that can be used as fuel because its atoms can be split.

fission: Splitting of an atom.

flexible fuel vehicle (FFV): A vehicle that can run on a variety of fuel types without modification of the engine.

flow: The volume of water in a river or stream, usually expressed as gallons or cubic meters per unit of time, such as a minute or second.

fluorescent lightbulb: A lightbulb that produces light not with intense heat but by exciting the atoms in a phosphor coating inside the bulb.

fossil fuel: An organic fuel made through the compression and heating of plant matter over millions of years, such as coal, petroleum, and natural gas.

fusion: The process by which the nuclei of light atoms join, releasing energy.

G

gas: An air-like substance that expands to fill whatever container holds it, including natural gas and other gases commonly found with liquid petroleum.

gasification: A process of converting the energy from a solid, such as coal, into gas.

gasohol: A blend of gasoline and ethanol.

gasoline: Refined liquid petroleum most commonly used as fuel in internal combustion engines.

geothermal: Describing energy that is found in the hot spots under the Earth; describing energy that is made from heat.

geothermal reservoir: A pocket of hot water contained within the Earth's mantle.

global warming: A phenomenon in which the average temperature of the Earth rises, melting icecaps, raising sea levels, and causing other environmental problems.

gradient: A gradual change in something over a specific distance.

green building: Any building constructed with materials that require less energy to produce and that save energy during the building's operation.

greenhouse effect: A phenomenon in which gases in the Earth's atmosphere prevent the sun's radiation from being reflected back into space, raising the surface temperature of the Earth.

greenhouse gas: A gas, such as carbon dioxide or methane, that is added to the Earth's atmosphere by human actions. These gases trap heat and contribute to global warming.

H

halogen lamp: An incandescent lightbulb that produces more light because it produces more heat, but lasts longer because the filament is enclosed in quartz.

Heisenberg uncertainty principle: The principle that it is impossible to know simultaneously both the location and momentum of a subatomic particle.

heliostat: A mirror that reflects the sun in a constant direction.

hybrid vehicle: Any vehicle that is powered in a combination of two ways; usually refers to vehicles powered by an internal combustion engine and an electric motor.

hybridized: The bringing together of two different types of technology.

hydraulic energy: The kinetic energy contained in water.

hydrocarbon: A substance composed of the elements hydrogen and carbon, such as coal, petroleum, and natural gas.

hydroelectric: Describing electric energy made by the movement of water.

hydropower: Any form of power derived from water.

I

implement: To put something into practice.

incandescent lightbulb: A conventional lightbulb that produces light by heating a filament to high temperatures.

infrastructure: The framework that is necessary to the functioning of a structure; for example, roads and power lines form part of the infrastructure of a city.

inlet: An opening through which liquid enters a device, or place.

internal combustion engine: The type of engine in which the burning that generates power takes place inside the engine.

isotope: A "species" of an element whose nucleus contains more neutrons than other species of the same element.

K

kilowatt-hour: One kilowatt of electricity consumed over a one-hour period.

kinetic energy: The energy associated with movement, such as water that is in motion.

Kyoto Protocol: An international agreement among many nations setting limits on emissions of greenhouse gases; intended to slow or prevent global warming.

L

lava: Molten rock contained within the Earth that emerges from cracks in the Earth's crust, such as volcanoes.

lift: The aerodynamic force that operates perpendicular to the wind, owing to differences in air pressure on either side of a turbine blade.

lignite: A soft brown coal with visible traces of plant matter in it that burns with a great deal of smoke and produces less heat than anthracite or bituminous coal.

liquefaction: The process of turning a gas or solid into a liquid.

LNG (liquefied natural gas): Gas that has been turned into liquid through the application of pressure and cold.

LPG (liquefied petroleum gas): A gas, mainly propane or butane, that has been turned into liquid through the use of pressure and cold.

lumen: A measure of the amount of light, defined as the amount of light produced by one candle.

M

magma: Liquid rock within the mantle.

magnetic levitation: The process of using the attractive and repulsive forces of magnetism to move objects such as trains.

mantle: The layer of the Earth between the core and the crust.

mechanical energy: The energy output of tools or machinery.

meltdown: Term used to refer to the possibility that a nuclear reactor could become so overheated that it would melt into the earth below.

mica: A type of shiny silica mineral usually found in certain types of rocks.

modular: An object which can be easily arranged, rearranged, replaced, or interchanged with similar objects.

mousse: A frothy mixture of oil and seawater in the area where an oil spill has occurred.

N

nacelle: The part of a wind turbine that houses the gearbox, generator, and other components.

natural gas: A gaseous hydrocarbon commonly found with petroleum.

negligible: To be so small as to be insignificant.

neutron: A particle with no electrical charge found in the nucleus of most atoms.

NGL (natural gas liquid): The liquid form of gases commonly found with natural gas, such as propane, butane, and ethane.

nonrenewable: To be limited in quantity and unable to be replaced.

nucleus: The center of an atom, containing protons and in the case of most elements, neutrons.

O

ocean thermal energy conversion (OTEC): The process of converting the heat contained in the oceans' water into electrical energy.

octane rating: The measure of how much a fuel can be compressed before it spontaneously ignites.

off-peak: Describing period of time when energy is being delivered at well below the maximum amount of demand, often nighttime.

oil: Liquid petroleum; a substance refined from petroleum used as a lubricant.

organic: Related to or derived from living matter, such as plants or animals; composed mainly of carbon atoms.

overburden: The dirt and rocks covering a deposit of coal or other fossil fuel.

oxygenate: A substance that increases the oxygen level in another substance.

ozone: A molecule consisting of three atoms of oxygen, naturally produced in the Earth's atmosphere; ozone is toxic to humans.

P

parabolic: Shaped like a parabola, which is a certain type of curve.

paraffin: A kind of alkane hydrocarbon that exists as a white, waxy solid at room temperature and can be used as fuel or as a wax for purposes such as sealing jars or making candles.

passive: A device that takes advantage of the sun's heat but does not use an additional source of energy.

peat: A brown substance composed of compressed plant matter and found in boggy areas; peat can be used as fuel itself, or turns into coal if compressed for long enough.

perpetual motion: The power of a machine to run indefinitely without any energy input.

petrochemicals: Chemical compounds that form in rocks, such as petroleum and coal.

petrodiesel: Diesel fuel made from petroleum.

petroleum: Liquid hydrocarbon found underground that can be refined into gasoline, diesel fuel, oils, kerosene, and other products.

pile: A mass of radioactive material in a nuclear reactor.

plutonium: A highly toxic element that can be used as fuel in nuclear reactors.

polymer: A compound, either synthetic or natural, that is made of many large molecules. These molecules are made from smaller, identical molecules that are chemically bonded.

pristine: Not changed by human hands; in its original condition.

productivity: The output of labor per amount of work.

proponent: Someone who supports an idea or cause.

proton: A positively charged particle found in the nucleus of an atom.

R

radioactive: Term used to describe any substance that decays over time by giving off subatomic particles such as neutrons.

RFG (reformulated gasoline): Gasoline that has an oxygenate or other additive added to it to decrease emissions and improve performance.

rem: An abbreviation for "roentgen equivalent man," referring to a dose of radiation that will cause the same biological effect (on a "man") as one roentgen of X-rays or gamma rays.

reservoir: A geologic formation that can contain liquid petroleum and natural gas.

reservoir rock: Porous rock, such as limestone or sandstone, that can hold accumulations of petroleum or natural gas.

retrofit: To change something, like a home, after it is built.

rotor: The hub to which the blades of a wind turbine are connected; sometimes used to refer to the rotor itself and the blades as a single unit.

S

scupper: An opening that allows a liquid to drain.

seam: A deposit of coal in the ground.

sedimentary rock: A rock formed through years of minerals accumulating and being compressed.

seismology: The study of movement within the earth, such as earthquakes and the eruption of volcanoes.

sick building syndrome: The tendency of buildings that are poorly ventilated, lighted, and humidified, and that are made with certain synthetic materials to cause the occupants to feel ill.

smog: Air pollution composed of particles mixed with smoke, fog, or haze in the air.

stall: The loss of lift that occurs when a wing presents too steep an angle to the wind and low pressure along the upper surface of the wing decreases.

strip mining: A form of mining that involves removing earth and rocks by bulldozer to retrieve the minerals beneath them.

stored energy: The energy contained in water that is stored in a tank or held back behind a dam in a reservoir.

subsidence: The collapse of earth above an empty mine, resulting in a damaged landscape.

surcharge: An additional charge over and above the original cost.

superconductivity: The disappearance of electrical resistance in a substance such as some metals at very low temperatures.

T

thermal energy: Any form of energy in the form of heat; used in reference to heat in the oceans' waters.

thermal gradient: The differences in temperature between different layers of the oceans.

thermal mass: The measure of the amount of heat a substance can hold.

thermodynamics: The branch of physics that deals with the mechanical actions or relations of heat.

tokamak: An acronym for the Russian-built toroidal magnetic chamber, a device for containing a fusion reaction.

transitioning: Changing from one position or state to another.

transparent: So clear that light can pass through without distortion.

trap: A reservoir or area within Earth's crust made of nonporous rock that can contain liquids or gases, such as water, petroleum, and natural gas.

trawler: A large commercial fishing boat.

Trombé wall: An exterior wall that conserves energy by trapping heat between glazing and a thermal mass, then venting it into the living area.

turbine: A device that spins to produce electricity.

U

uranium: A heavy element that is the chief source of fuel for nuclear reactors.

V

viable: To be possible; to be able to grow or develop.

voltage: Electric potential that is measured in volts.

W

wind farm: A group of wind turbines that provide electricity for commercial uses.

work: The conversion of one form of energy into another, such as the conversion of the kinetic energy of water into mechanical energy used to perform a task.

Z

zero point energy: The energy contained in electromagnetic fluctuations that remains in a vacuum, even when the temperature has been reduced to very low levels.

Overview

In the technological world of the twenty-first century, few people can truly imagine the challenges faced by prehistoric people as they tried to cope with their natural environment. Thousands of years ago life was a daily struggle to find, store, and cook food, stay warm and clothed, and generally survive to an "old age" equal to that of most of today's college students. A common image of prehistoric life is that of dirty and ill-clad people huddled around a smoky campfire outside a cave in an ongoing effort to stay warm and dry and to stop the rumbling in their bellies.

The "caves" of the twenty-first century are a little cozier. The typical person, at least in more developed countries, wakes up each morning in a reasonably comfortable house because the gas, propane, or electric heating system (or electric air-conditioner) has operated automatically overnight. A warm shower awaits because of hot water heaters powered by electricity or natural gas, and hair dries quickly (and stylishly) under an electric hair dryer. An electric iron takes the wrinkles out of the clean shirt that sat overnight in the electric clothes dryer. Milk for a morning bowl of cereal remains fresh in an electric refrigerator, and it costs pennies per bowl thanks to electrically powered milking operations on modern dairy farms. The person then goes to the garage (after turning off all the electric lights in the house), hits the electric garage door opener, and gets into his or her gasoline-powered car for the drive to work—perhaps in an office building that consumes power for lighting, heating and air-conditioning, copiers, coffeemakers, and computers. Later, an electric, propane, or natural gas stove is used to cook dinner. Later still, an electric popcorn popper provides a snack as the person watches an electric television or reads under the warm glow of electric light bulbs—after perhaps turning up the heat because the house is a little chilly.

CATASTROPHE AHEAD?

Most people take these modern conveniences for granted. Few people give much thought to them, at least until there is a power outage or prices rise sharply, as they did for gasoline in the United States in the summer and fall of 2005. Many scientists, environmentalists, and concerned members of the public, though, believe that these conveniences have been taken too much for granted. Some believe that the modern reliance on fossil fuels—fuels such as natural gas, gasoline, propane, and coal that are processed from materials mined from the earth—has set the Earth on a collision course with disaster in the twenty-first century. Their belief is that the human community is simply burning too much fuel and that the consequences of doing so will be dire (terrible). Some of their concerns include the following:

- Too much money is spent on fossil fuels. In the United States, over $1 billion is spent every day to power the country's cars and trucks.
- Much of the supply of fossil fuels, particularly petroleum, comes from areas of the world that may be unstable. The U.S. fuel supply could be cut off without warning by a foreign government. Many nations that import all or most of their petroleum feel as if they are hostages to the nations that control the world's petroleum supplies.
- Drilling for oil and mining coal can do damage to the landscape that is impossible to repair.
- Reserves of coals and especially oil are limited, and eventually supplies will run out. In the meantime, the cost of such fuels will rise dramatically as it becomes more and more difficult to find and extract them.
- Transporting petroleum in massive tankers at sea heightens the risk of oil spills, causing damage to the marine and coastal environments.

Furthermore, to provide heat and electricity, fossil fuels have to be burned, and this burning gives rise to a host of problems. It releases pollutants in the form of carbon dioxide and sulfur into the air, fouling the atmosphere and causing "brown clouds" over cities. These pollutants can increase health problems such as lung disease. They may also contribute to a phenomenon called "global warming." This term refers to the theory that average temperatures across the globe will increase as "greenhouse gases" such as carbon dioxide trap the sun's heat (as a greenhouse does) in the atmosphere and warm it. Global warming, in turn, can melt glaciers and the polar ice caps, raising sea levels with damaging effects on coastal cities and small island nations. It may also cause climate changes, crop failures, and more unpredictable weather patterns.

Some scientists do not believe that global warming even exists or that its consequences will be catastrophic. Some note that throughout history, the world's average temperatures have risen and fallen. Some do not find the scientific data about temperature, glacial melting, rising sea levels, and unpredictable weather totally believable. While the debate continues, scientists struggle to learn more about the effects of human activity on the environment. At the same time, governments struggle to maintain a balance between economic development and its possible effects on the environment.

WHAT TO DO?

These problems began to become more serious after the Industrial Revolution of the nineteenth century. Until that time people depended on other sources of power. Of course, they burned coal or wood in fireplaces and stoves, but they also relied on the power of the sun, the wind, and river currents to accomplish much of their work. The Industrial Revolution changed that. Now, coal was being burned in vast amounts to power factories and steam engines as the economies of Europe and North America grew and developed. Later, more efficient electricity became the preferred power source, but coal still had to be burned to produce electricity in large power plants. Then in 1886 the first internal combustion engine was developed and used in an automobile. Within a few decades there was a demand for gasoline to power these engines. By 1929 the number of cars in the United States had grown to twenty-three million, and in the quarter-century between 1904 and 1929, the number of trucks grew from just seven hundred to 3.4 million.

At the same time technological advances improved life in the home. In 1920, for example, the United States produced a total of five thousand refrigerators. Just ten years later the number had grown to one million per year. These and many other industrial and consumer developments required vast and growing amounts of fuel. Compounding the problem in the twenty-first century is that other nations of the world, such as China and India, have started to develop more modern industrialized economies powered by fossil fuels.

By the end of World War II in 1945, scientists were beginning to imagine a world powered by fuel that was cheap, clean, and inexhaustible (unable to be used up). During the war the United States had unleashed the power of the atom to create the atomic bomb. Scientists believed that the atom could be used for peaceful purposes in nuclear power plants. They even envisioned (imagined) a day when homes could be powered by their own tiny nuclear power generators. This dream proved to be just that. While some four hundred nuclear power plants worldwide provide about 16 percent of the world's electricity, building such plants is an enormously expensive technical feat. Moreover, nuclear power plants produce spent fuel that is dangerous and not easily disposed of. The public fears that an accident at such a plant could release deadly radiation that would have disastrous effects on the surrounding area. Nuclear power has strong defenders, but it is not cheap, and safety concerns sometimes make it unpopular.

The dream of a fuel source that is safe, plentiful, clean, and inexpensive, however, lives on. The awareness of the need for such alternative fuel sources became greater in the 1970s, when the oil-exporting countries of the Middle East stopped shipments of oil to the United States and its allies. This situation (an embargo) caused fuel shortages and rapidly rising prices at the gas pump. In the decades that followed, gasoline again became plentiful and relatively inexpensive, but the oil embargo served as a wakeup call for many people. In addition, during these years people worldwide grew concerned about pollution, industrialization, and damage to the environment. Accordingly, efforts were intensified to find and develop alternative sources of energy.

ALTERNATIVE ENERGY: BACK TO THE FUTURE

Some of these alternative fuel sources are by no means new. For centuries people have harnessed the power of running water for a variety of needs, particularly for agriculture (farming). Water wheels were constructed in the Middle East, Greece, and China thousands of years ago, and they were common fixtures on the farms of Europe by the Middle Ages. In the early twenty-first century hydroelectric dams, which generate electricity from the power of rivers, provide about 9 percent of the electricity in the

United States. Worldwide, there are about 40,000 such dams. In some countries, such as Norway, hydroelectric dams provide virtually 100 percent of the nation's electrical needs. Scientists, though, express concerns about the impact such dams have on the natural environment.

Water can provide power in other ways. Scientists have been attempting to harness the enormous power contained in ocean waves, tides, and currents. Furthermore, they note that the oceans absorb enormous amounts of energy from the sun, and they hope someday to be able to tap into that energy for human needs. Technical problems continue to occur. It remains likely that ocean power will serve only to supplement (add to) existing power sources in the near future.

Another source of energy that is not new is solar power. For centuries, people have used the heat of the sun to warm houses, dry laundry, and preserve food. In the twenty-first century such "passive" uses of the sun's rays have been supplemented with photovoltaic devices that convert the energy of the sun into electricity. Solar power, though, is limited geographically to regions of the Earth where sunshine is plentiful.

Another old source of heat is geothermal power, referring to the heat that seeps out of the earth in places such as hot springs. In the past this heat was used directly, but in the modern world it is also used indirectly to produce electricity. In 1999 over 8,000 megawatts (that is, 8,000 million watts) of electricity were produced by about 250 geothermal power plants in twenty-two countries around the world. That same year the United States produced nearly 3,000 megawatts of geothermal electricity, more than twice the amount of power generated by wind and solar power. Geothermal power, though, is restricted by the limited number of suitable sites for tapping it.

Finally, wind power is getting a closer look. For centuries people have harnessed the power of the wind to turn windmills, using the energy to accomplish work. In the United States, wind-operated turbines produce just 0.4 percent of the nation's energy needs. However, wind experts believe that a realistic goal is for wind to supply 20 percent of the nation's electricity requirements by 2020. Worldwide, wind supplies enough power for about nine million homes. Its future development, though, is hampered by limitations on the number of sites with enough wind and by concerns about large numbers of unsightly wind turbines marring the landscape.

ALTERNATIVE ENERGY: FORWARD TO THE FUTURE

While some forms of modern alternative energy sources are really developments of long-existing technologies, others are genuinely new, though scientists have been exploring even some of these for up to hundreds of years. One, called bioenergy, refers to the burning of biological materials that otherwise might have just been thrown away or never grown in the first place. These include animal waste, garbage, straw, wood by-products, charcoal, dried plants, nutshells, and the material left over after the processing of certain foods, such as sugar and orange juice. Bioenergy also includes methane gas given off by garbage as it decomposes or rots. Fuels made from vegetable oils can be used to power engines, such as those in cars and trucks. Biofuels are generally cleaner than fossil fuels, so they do not pollute as much, and they are renewable. They remain expensive, and amassing significant amounts of biofuels requires a large commitment of agricultural resources such as farmland.

Nothing is sophisticated about burning garbage. A more sophisticated modern alternative is hydrogen, the most abundant element in the universe. Hydrogen in its pure form is extremely flammable. The problem with using hydrogen as a fuel is separating hydrogen molecules from the other elements to which it readily bonds, such as oxygen (hydrogen and oxygen combine to form water). Hydrogen can be used in fuel cells, where water is broken down into its elements. The hydrogen becomes fuel, while the "waste product" is oxygen. Many scientists regard hydrogen fuel cells as the "fuel of the future," believing that it will provide clean, safe, renewable fuel to power homes, office buildings, and even cars and trucks. However, fuel cells are expensive. As of 2002 a fuel cell could cost anywhere from $500 to $2,500 per kilowatt produced. Engines that burn gasoline cost only about $30 to $35 for the same amount of energy.

All of these power sources have high costs, both for the fuel and for the technology needed to use it. The real dreamers among energy researchers are those who envision a future powered by a fuel that is not only clean, safe, and renewable but essentially free. Many scientists believe that such fuel alternatives are impossible, at least for the foreseeable future. Others, though, work in laboratories around the world to harness more theoretical sources of energy. Some of their work has a "science fiction" quality, but these scientists point out that a few hundred years ago the airplane was science fiction.

One of these energy sources is magnetism, already used to power magnetic levitation ("maglev") trains in Japan and Germany. Another is perpetual motion, the movement of a machine that produces energy without requiring energy to be put into the system. Most scientists, though, dismiss perpetual motion as a violation of the laws of physics. Other scientists are investigating so-called zero-point energy, or the energy that surrounds all matter and can even be found in the vacuum of space. But perhaps the most sought-after source of energy for investigators is cold fusion, a nuclear reaction using "heavy hydrogen," an abundant element in seawater, as fuel. With cold fusion, power could be produced literally from a bucket of water. So far, no one has been able to produce it, though some scientists claim to have come very close.

None of these energy sources is a complete cure for the world's energy woes. Most will continue to serve as supplements to conventional fossil fuel burning for decades to come. But with the commitment of research dollars, it is possible that future generations will be able to generate all their power needs in ways that scientists have not even yet imagined. The first step begins with understanding fossil fuels, the energy they provide, the problems they cause, and what it may take to replace them.

Hydrogen

INTRODUCTION: WHAT IS HYDROGEN ENERGY?

Hydrogen, the first element in the periodic table, is one of the most common elements found on Earth and the lightest one known to exist. An estimated 90 percent of the universe is composed of hydrogen. It can be found in nearly everything organic (that is, any material that contains the element carbon except diamond and graphite) and in all living organisms. In its pure gaseous form, hydrogen is odorless, colorless, tasteless, highly flammable, but not poisonous.

Many experts believe that hydrogen could be used as a fuel source to provide energy to the world. In order for this to happen, the gas must be in its pure form. This is problematic because hydrogen bonds (connects or attaches) relatively easily to other elements. In fact, it does not occur as a gas in nature but rather is found in combination with other elements. For example, hydrogen combines with oxygen to form water. Because water is so common, most methods to produce hydrogen gas focus on extracting it from water.

Electrolysis, a process that uses electricity, can separate the hydrogen from the oxygen in water. Photolysis detaches the elements from each other using sunlight instead of produced electricity. It is also possible to make the hydrogen industrially, by using methods such as steam reformation. In all cases, isolating the hydrogen yields a gas that is suitable for use as a fuel source.

Once the hydrogen is in pure form, it can be used several different ways. One use is to make a hydrogen fuel cell that can be used to power electrical generators or vehicles. Another is to

Words to Know

Conductive A material that can transmit electrical energy.

Electrolysis A method of producing chemical energy by passing an electric current through a type of liquid.

Emission The release of substances into the atmosphere. These substances can be gases, greenhouse gases, or particles.

Geothermal Describing energy that is found in the hot spots under the Earth; describing energy that is made from heat.

Greenhouse gas A gas, such as carbon dioxide or methane, that is added to the Earth's atmosphere by human actions. These gases trap heat and contribute to global warming.

Infrastructure The underlying foundation or basic framework of a system, such as buildings or equipment.

Off-peak Describing periods of time when energy is being delivered at well below the maximum amount of demand, often nights.

use hydrogen to power an internal combustion engine (ICE), just like the ICEs that are already used to power cars and other vehicles. Using hydrogen in these ways can have both benefits and drawbacks, all of which are related to economical, societal, and environmental circumstances present in today's world.

HISTORICAL OVERVIEW

The use of hydrogen as a fuel source is not a modern notion. Scientists and visionaries have been experimenting with hydrogen since the seventeenth century. Its potential is still being explored in the twenty-first century.

Finding hydrogen

Hydrogen was first produced as early as 1671, when Robert Boyle (1627–1691), an English chemist, dissolved (mixed or melted) iron in acid. Boyle and other early scientists were unaware that hydrogen was a unique element. In fact, it was not until 1766 that hydrogen was officially recognized as an individual gas. Another English chemist, Henry Cavendish (1731–1810), measured the density of several gases to prove that hydrogen existed. He found that hydrogen was almost fourteen times lighter than ordinary air and called it "inflammable air" (meaning air that is likely to burn or explode).

Following Cavendish's lead, a French scientist named Antoine-Laurent Lavoisier (1743–1794) repeated Cavendish's experiments in 1785 and gave hydrogen its name, from the Greek words *hydro*, meaning water, and *genes*, meaning forming. In addition,

This 18th century engraving shows four men filling a hydrogen balloon in Paris. The gas was produced by pouring sulfuric acid upon filings of iron. © *UPI/Corbis-Bettman.*

Lavoisier's process for isolating hydrogen (a rudimentary form of electrolysis) became the primary method for obtaining hydrogen gas up through the early nineteenth century.

Hot Air or Hydrogen?

There is often confusion between the first hot air balloon flights and the first hydrogen balloon flights. Hot air balloon flights also originated in France but predated hydrogen flights by only a few months. Two Frenchmen, Joseph (1740–1810) and Étienne (1745–1799) Montgolfier, built a hot air balloon big enough to carry a basket, which in turn carried a duck, a sheep, and a rooster. This balloon's first flight occurred on September 19, 1783, only a few months before Jacques Charles's December flight that same year. The Montgolfier brothers went on to build several hot air balloons, one of which still holds a record as one of the largest balloons ever made. The balloon was flown by Joseph Montgolfier himself in 1784.

After the Montgolfiers' first flight, another Frenchman, Jean Blanchard (1753–1809), and John Jeffries, an American doctor from Boston, crossed the English Channel in a hot air balloon in 1785. Blanchard is also credited with the first hot air balloon flights in Germany, Poland, and the Netherlands. In 1793 Blanchard made a flight from Philadelphia, Pennsylvania, to New Jersey and delivered a letter, which became the first piece of airmail to travel in the United States. The ascent was witnessed by President George Washington, who with other onlookers, had paid Blanchard for the privilege.

Hydrogen balloon history

The history of hydrogen balloon flight began in France in December 1783, with the French physicist Jacques Charles (1746–1823). Charles and a companion, Noel Roberts, who helped build the balloon, were the first people ever to ascend in a hydrogen-filled balloon. They traveled 27 miles (43 kilometers) before the balloon came safely to rest. Charles is credited with the first solitary hydrogen balloon flight, during which he rose up 10,000 feet (3 kilometers) before landing again.

The first hydrogen fuel cell

In 1839 Sir William Grove (1811–1896) built the first working fuel cell. Grove, an amateur scientist and a Welsh judge, was

The 2005 Honda FCX fuel cell powered vehicle is seen on display during its launch at the Petersen Automotive Museum in Los Angeles on June 29, 2005. © *Mario Anzuoni/Reuters/Corbis.*

aware that an electric current (the movement or flow of electrons) could split a molecule of water into its component parts, hydrogen and oxygen, in a process known as electrolysis. He therefore deduced that, under the right circumstances, he might be able to produce water and electricity by combining hydrogen and oxygen. Grove conducted his experiment by putting strips of platinum into two different bottles, one full of hydrogen and one full of oxygen. He then placed the bottles into an electrolyte (a chemical substance that is capable of conducting current), in this case, sulfuric acid, where current began to flow and water accumulated in the gas bottles. Although Grove's fuel cell did work, he never found a practical use for it, and he never named it. Two chemists, Ludwig Mond and Charles Langer, coined the term *fuel cell* in 1889.

Moving on to airships

Airships were introduced in the nineteenth century and became another means of transportation that used hydrogen as a fuel source. Also known as a dirigible, an airship differs from a hydrogen balloon because it has a steering mechanism, often including an engine of some kind. There are three types of airships: a nonrigid airship, or a blimp; a semirigid airship, and a

What's the Difference Between a Fuel Cell and a Battery?

A battery and a fuel cell are both electrochemical devices that convert chemical energy into electrical energy. The chemical reaction in a battery releases electrons that travel between the terminals and out as electricity. Moreover, when electricity is released from the battery, the battery's stored energy is being used up because the battery is a closed storage system. It can only produce so much energy before it dies and needs to be recharged or replaced. The fuel cell, on the other hand, is more of an energy converter than an energy storage device. Its chemical reaction converts hydrogen and oxygen into water and in the process produces electricity. A fuel cell will provide power as long as it is supplied with fuel. It does not run down or require recharging like a battery. A fuel cell can be refilled with hydrogen like filling an automobile gas tank.

rigid airship (dirigible) or zeppelin, named after the first to build them, Count Ferdinand Adolf August Heinrich Zeppelin. All airships are sometimes known as LTA craft because the gas that provides their lift is lighter than air.

In the early twentieth century airships were used by the militaries of countries such as Germany and Great Britain. Airships also were sometimes used to carry passengers for long-distance travel. When airships were used as a means of transportation, they were often luxurious and expensive. Passengers sometimes boarded the airships to travel across the ocean. When traveling from Europe, for example, a person could reach the United States more quickly than by ocean liner.

One innovative airship that used hydrogen as the means of inflation was called the *Akron*. It was built in 1911 by Melvin Vaniman (1866–1912). The engine that powered the *Akron* could be run on gasoline or hydrogen. A flick of a lever changed which fuel was being used. Unfortunately, the *Akron* never got much use as a passenger carrier.

Germany built the greatest number of hydrogen-filled airships. Some of these airships even traveled around the globe. One of the best known zeppelins was the *Graf Zeppelin*. It began running in 1928 and went around the world twice in 1929 alone. Over its ten-year active lifespan, the *Graf Zeppelin* traveled over one million miles (1,609,344 kilometers). It had no accidents, unlike many other hydrogen airships. In 1937 Hydrogen developed a negative reputation because of a disaster involving another German airship, the *Hindenburg*. International law now bans the use of hydrogen as an inflating gas for airships.

Syngas

Vehicles were not the only use of hydrogen in the late nineteenth and early twentieth centuries. Hydrogen is part of a fuel called syngas, which is also known as synthetic gas or town gas. Syngas is made up of as much as 50 percent hydrogen. It is made from coal, wood, and some waste that has been gasified (made into a gas). In the United States, syngas was first used as early as the late 1700s. It became a more common fuel in the late nineteenth century and until about 1940. Primarily used in urban areas to provide a fuel for heat and for cooking, it was also used in Europe and other parts of the world in the same time period. In Europe, syngas provided light for city streets, homes, and public buildings. It is still used in parts of China, Europe, and South America, where natural gas is not a fueling option.

Other twentieth-century research developments

Though some work on hydrogen as a fuel source was done in the nineteenth century, more work was done in the first half of the twentieth century. In the 1920s and 1930s European scientists and engineers experimented with the use of hydrogen as a fuel. Among their accomplishments was converting several types of vehicles to run on hydrogen, including trucks, a bus, and a railcar that was self-propelled.

In planes and space

Hydrogen did find some uses in aviation and the space program in this time period. Hydrogen was used to fuel a jet engine as early as the late 1950s on an experimental basis. By the late 1980s more research was being conducted in the United States and Russia in the use of plane engines fueled by hydrogen. Some supersonic jets might use hydrogen in the future, if the technology can be developed.

NASA has used hydrogen in various capacities since the 1950s. Hydrogen fuel cells provided power for the manned Gemini and Apollo space flights in the 1960s and 1970s. Fuel cells were used on these craft because they were seen as safer

The Graf Zeppelin approaching the mooring mast at Mines Field (Los Angeles) after completing its trip from Tokyo in 68 hours for the third successful lap of its historic round the world flight. © *Bettmann/Corbis.*

than nuclear power, another option that was considered. Another benefit of using hydrogen fuel cells on these flights was that the by-product of fuel cells—water—could be consumed by the astronauts. Liquid hydrogen has also been used in the space program as a rocket fuel to propel vehicles into space. In addition, space shuttles run by NASA since the 1980s have employed hydrogen as a fuel.

This use of hydrogen led to a tragedy. When a rubber seal failed on the space shuttle *Challenger* as it was lifting off in 1986, hydrogen gas mixed with the flame that was propelling the rocket *Challenger* into space. The mixture caused the space shuttle to explode. There were seven astronauts aboard, all of whom lost their lives.

First hydrogen research organization

There was continued interest in hydrogen as a fuel for other uses in the 1960s and 1970s. In the mid-1970s the modern era of

The *Hindenburg* Tragedy

In 1937, the German dirigible *LZ 129*, nicknamed the *Hindenburg*, traveled from Germany to the United States with a number of passengers. Including the crew, about 97 people were aboard. When the *Hindenburg* reached Lakehurst, New Jersey, the ship exploded, killing 36 people. Only 13 were passengers. The rest were crew members and one American who was on the ground at the time of the explosion. The investigation into the incident concluded that the hydrogen inside the dirigible probably caused the explosion. Investigators in the 1930s believed that electric discharge from the atmosphere ignited the hydrogen. Because of these findings, hydrogen began developing a negative reputation in the general public's mind.

This reputation was not deserved. Many years later, a scientist named Addison Bain (1935–), who worked for NASA (the National Aeronautics and Space Administration) as manager of its hydrogen program, investigated the *Hindenburg* tragedy. He believed that the *Hindenburg* accident was not caused by the hydrogen exploding. He noted that the outer shell of the dirigible was a cotton cover that was painted with some flammable chemicals to both decorate and reinforce the airship's shell. Bain believed the substances were ignited by the static charges that had built up on the ship's metal frame as a result of a very stormy environment. What had been painted on the dirigible acted like rocket fuel. The resulting explosion caused the disaster.

Bain concluded that the flame color also revealed that the fire could not have been started by the hydrogen. Witnesses from 1937 reported that the flames were colorful. However, hydrogen burns almost clear

hydrogen research began. In this phase, hydrogen was regarded as an energy source to replace fossil fuels. The first international conference was held in Miami Beach, Florida, and was called the Hydrogen Economy Miami Energy Conference. This event led to the founding of the International Association for Hydrogen Energy, an organization that in the 1990s helped get research off the ground and led to a growth of organizations, studies, and research all focused on hydrogen energy.

Twenty-first century developments

Several countries have put much effort into the study, support, and use of hydrogen as an alternative fuel for the future, including Canada, Japan, Germany, and the United States. Each country has its own vision, but most have pledged at least some public funding. The European Union has also pledged to spend money to help create hydrogen fuel cells through a partnership between

The *Hindenburg* blimp, crashing into metal structure, with its tail and more than one third of body in flames, May 6, 1937. © *Hindenburg, May 6, 1937.*

in the daylight, the time when the incident took place. Despite Bain's findings, many people still believe that the hydrogen exploded and caused the disaster.

government and business. One country in particular, Iceland, has already committed to replacing its oil imports with hydrogen-fueled technology and is currently one of the largest consumers of hydrogen fuel.

Research in the United States

Most vehicles on the road today are powered by gasoline, which is produced from oil. Because oil will eventually run out, alternatives are needed to fuel vehicles in the future. A significant amount of money from both private and public sources is being invested in the early twenty-first century to develop hydrogen technology for vehicles in the United States. The concentrated movement to embrace hydrogen as an alternative energy began in 1990 with the passage of the federal Clean Air Act. This act called for a reduction in air pollution by changing the design of cars. The act also sought to change the kind of fuels that cars used so that their emissions (the waste by-product that is expelled by each vehicle) would be reduced. In addition, new emission standards were called for. Though hydrogen and other alternative fuels were not named specifically, hydrogen was a technology that was explored as a possible means of meeting this act's goal.

After the passage of the Clean Air Act, California was one state that pursued alternative energy technologies, including hydrogen. The state was especially interested in alternative fuels because the state had a major problem with air pollution. In California, which had about 30 million vehicles on the road as of 2005, about 90 percent of the population live where air quality cannot meet federal standards. California has addressed this problem in several ways. For example, some of the toughest standards for emissions in the United States can be found in California. Another way is through the work of the California Fuel Cell Partnership. This is a group dedicated to making fuel cells and vehicles that run on fuel cells part of American life. The partnership includes the government, companies that make fuel cells, energy providers, and car companies. In addition to educating the public about hydrogen fuel cell technology, the partnership works toward getting hydrogen fuel cell cars on the road and making hydrogen fuel stations available. By 2007 the partnership hopes to have 300 hydrogen fuel cell cars and buses on the road.

In 2002 and 2003 the United States made a significant commitment to embracing hydrogen in the form of fuel cell technology. In 2002, Secretary of Energy Spencer Abraham announced an initiative called FreedomCAR. A partnership between the federal government and U.S. car makers, this initiative pushed for research on hydrogen fuel cell technology. About $500 million was to be spent on this proposal.

President George W. Bush (1946–) built on the proposal in his January 2003 State of the Union address. The president's proposal, called the FreedomCAR and Fuel Initiative, included spending $1.2 billion over five years in research conducted by both the government and private companies, such as car manufacturers, refineries, and chemical companies. The funds were designed to help create fuel cell technology for cars and trucks as well as homes and businesses. The hydrogen to power these cells would be created through electricity production, primarily from next-generation nuclear power plants and electric plants that run on coal. About $720 million of the funds were to go to building the infrastructure (the basic facilities, services and installations) needed to make the hydrogen, store it, and distribute it. Funds

The space shuttle *Challenger* exploding shortly after lifting off from Kennedy Space Center. *AP Images.*

were included specifically to develop new technologies for cars, a significant issue in using hydrogen as a fuel source.

The federal government had a stated goal of putting hydrogen fuel cell cars on U.S. roads by 2010. The government hoped that hydrogen fuel cell cars would be the norm by 2020. The United States also supported the International Partnership for the Hydrogen Economy, which deals with the creation of the hydrogen economy on a worldwide basis. Some scientists and alternative energy supporters were critical of the proposal. Some were not pleased that other alternative energy sources did not receive money. Others were critical of the fact that the proposal still backed energy sources such as coal and nuclear power as the fuel to make the hydrogen. Coal, like oil, will one day run out, and many believe that hydrogen should be made from a renewable resource instead.

Japanese research

The Japanese government is very committed to developing hydrogen-based technologies because the country depends on foreign oil. The Japanese want to lessen or end their need for

imported oil through the development of alternative energy sources such as hydrogen. The Japanese government spends several hundred million dollars each year on research into hydrogen fuel and fuel cells. In 2004 alone, the Japanese government spent $268 million on fuel cell research and development.

The hydrogen genset is capable of producing 114 k VA of power at several voltage levels and is based upon a standard 6.8-liter Ford production engine that has been modified for hydrogen use. © *Reuters/Corbis.*

The Japanese government wants 50,000 cars powered by hydrogen fuel cells to be on the road by 2010. By 2020 the government wants the number to increase to five million. The government also hopes to have 4,000 hydrogen filling stations along Japanese roads by 2020.

Research in Canada and Germany

In the twentieth century Canada spent several decades researching fuel cells—not using hydrogen, but an alkaline electrolyte or phosphoric acid as an electrolyte. Beginning in 1980 and into the late 1990s, the country started to experiment with hydrogen fuel cells. One company, Stuart Energy, promised to build five stations where vehicles could obtain hydrogen fuel by 2005. The Canadian government has pledged $500 million over five years, in the first decade of the twenty-first century, for fuel cell research.

In the 1950s Germany did research into alkaline fuel cells, while hydrogen research blossomed later in the century. By 2003 over 350 groups in Germany were working on hydrogen fuel cell technology.

Commitment in Iceland

Iceland wants to be the first country whose energy system is based on hydrogen. Iceland is a small island of only 40,000 square miles (64,374 square kilometers) near the Arctic Circle. The country's population is fewer than 300,000 people. Iceland's limited space and population make it an ideal place to test whether a hydrogen economy will work. The country decided to embrace hydrogen before the end of the twentieth century, with the goal of being fully hydrogen-based by midway through the twenty-first century.

Icelanders want to be self-sufficient in terms of energy. The country is already capable of producing more than enough of its own energy for heating and cooling purposes. However, because its population uses cars, buses, and ships, Iceland must import oil. This oil accounts for 30 percent of the country's energy consumption. Iceland wants to reduce this figure to zero. To reach this goal, a joint venture company was created in the late 1990s. It is called Icelandic New Energy and includes input from companies including Shell Hydrogen, Norsky Hydro, and DaimlerChrysler. In 2000 the company began creating the infrastructure for production and distribution of hydrogen as fuel. Iceland has already decided that most of its hydrogen energy will come from fuel cells, which will be used in generators and vehicles.

By 2003 Iceland had its first hydrogen retail outlet, a Shell filling station, in its main city of Reykjavik. Hydrogen was produced on site using hydroelectric and geothermal energy to power the reaction. The hydrogen produced there was also being stored and distributed to other locations. Some of the first users of this hydrogen filling station were three public transit buses. These buses look like standard buses, but they are taller because the hydrogen tanks are located on the roof. Iceland has faced some problems with these buses. They must be kept inside at night so they keep warm. Officials do not want to have the water emitted by the fuel cells freeze and damage the cells. While the buses are being gradually introduced, Iceland next wants to get automobiles that run on hydrogen fuel cells to be the standard vehicle of choice. The country expects to introduce such cars in 2006.

Down the road, a bigger challenge will be getting boats and ships to run on hydrogen technology. Most of Iceland's fossil fuel consumption comes from the use of boats for fishing, a staple of the Icelandic economy. Powering boats with fuel cells is more challenging because a trawler (a boat designed to catch fish by dragging large nets), for example, carries a large amount of gasoline and stays at sea for several days. More hydrogen than that would be needed for a trip of the same length. The Icelandic government will have to convince those who use boats to accept hydrogen as a fuel. Iceland wants to run exclusively on hydrogen by 2050.

PRODUCING HYDROGEN

Hydrogen is sometimes considered to be the energy source of the future, for a few reasons. One reason for this belief is that hydrogen is renewable. Unlike the fossil fuels upon which the world is currently dependent, hydrogen can be produced or "created" and in a short amount of time. There are several methods by which hydrogen can be produced, including, but not limited to, electrolysis and steam reforming.

Electrolysis

Electrolysis is the process by which an electric current is passed through water and breaks the chemical bonds between hydrogen and oxygen. An electrolyte, a fluid chemical substance that can carry a current, aids in the bond-breaking procedure. Once the bonds are broken, the atomic components (hydrogen and oxygen) become either positive or negative ions (charged particles). Two terminals (anode and cathode) also have positive and negative charges, drawing the resulting ions toward them. Generally, the positive hydrogen ions gather at the anode (which is negative), while the negative oxygen ions reside at the cathode (which is positive). Gas is then formed at either terminal.

It is possible to perform electrolysis at high temperatures. High temperature electrolysis (HTE), also known as steam electrolysis, operates much the same way as conventional electrolysis. The variation occurs in that, rather than using a standard amount of electric current, heat is applied instead. This reduces the total amount of electric energy required to produce hydrogen gas.

Steam reforming

Steam reforming, sometimes called reforming or steam methane reforming, is another well-known method for making hydrogen. Natural gas is the most common fuel used in steam reforming. To make hydrogen using steam reforming, natural gas is reacted with steam at a very high temperature in a combustion chamber. The temperature can be from 1472°–3982°F (800°–1700°C).

A catalyst (a substance that increases the rate of a reaction without being consumed in the process) is present in some steam reformers. The catalyst is usually made of metal. The catalyst helps

A semiconductor is immersed in the water and splits water molecules using the energy in sunlight. The water molecules split into hydrogen and oxygen gas. Burning the hydrogen in oxygen releases the stored energy and reforms water, completing the cycle. *NREL/U.S. Department of Energy/Photo Researchers, Inc.*

break up the natural gas into methane. When the methane and water react, hydrogen is produced. Carbon oxides such as carbon monoxide and carbon dioxide are made as by-products. In some processes, the carbon monoxide is reacted again to form more hydrogen and carbon dioxide.

The steam reforming process has some positive points. Of all the fossil fuels, natural gas is the cleanest burning. In other words, it gives off fewer by-products that can contribute to pollution. The use of natural gas to make hydrogen might help in the creation of an infrastructure for the distribution of hydrogen. Since there are stations that already distribute natural gas, the natural gas could be transported there and converted to hydrogen via steam reforming on site and on a small scale. This means of production could provide hydrogen for cars that run on either hydrogen fuel cells or hydrogen-powered internal combustion engines.

Benefits and drawbacks of existing production methods

Each hydrogen-producing method has its own benefits and drawbacks. Electrolysis is considered to be the most environmentally

Other Production Methods

Scientists from around the world are trying to find the best way to make hydrogen from renewable resources and have come up with many unique ideas. For example, since the 1940s, scientists have worked to use algae (such as pond scum) to make hydrogen. Algae naturally produce hydrogen from water using sunlight energy, a process called photolysis. More recently, a scientist in England, Murat Dogru, proposed that hazelnuts could provide a source of hydrogen, because hazelnut shells produce hydrogen when they are burned.

Bacteria are also being investigated as a way to make hydrogen, but this is not commercially practical yet. Bacteria react like algae in water and can naturally separate the hydrogen and oxygen using sunlight. Experiments are being conducted to alter the structure of the bacteria so that they produce less oxygen and more hydrogen to be used as fuel. Another method of producing hydrogen employs microbes (microorganisms). These microbes are used to make biomass (the leftovers from crops that cannot be used anywhere else) into hydrogen.

Another potential innovation begins with biogas (containing methane, carbon dioxide, water vapor, and other gases) that is caught from the gaseous releases of dairy cows. The biogas is converted to hydrogen and used to power fuel cells. The fuel cells are intended for use in hydrogen-powered generators on the farms. In 2004 scientists working at the University of Minnesota, Twin Cities, discovered a way of taking corn, fermenting it, producing ethanol, and converting it into hydrogen fuel.

friendly procedure, because it produces no by-products that are harmful to the environment. In addition, it has a potentially positive by-product: oxygen. This oxygen could be captured and used elsewhere.

However, large-scale production of hydrogen by electrolysis becomes very expensive because electricity is used to create the electric currents. If renewable energy sources such as solar energy, hydropower, hydroelectric power, or even nuclear power were used to produce the current, the process would become much more affordable. Another source of energy could be obtained through the use of biomass: waste, sewage, and agricultural residue are all endlessly renewable and have little negative effect on the environment.

The steam reforming process is the most common method used to make hydrogen industrially. One benefit is that it is cheaper than producing hydrogen by electrolysis. However, a big drawback is the amount of carbon dioxide produced during the process. If the steam reforming process is to catch on as a means of mass-producing hydrogen fuel, the issue of what to do with the carbon dioxide produced must be addressed. Carbon dioxide can build up and trap heat on the planet. This condition is known as global warming. Potential solutions to the carbon dioxide issue with steam reforming exist, and all are costly. The carbon dioxide could be stored in empty gas wells or oil wells where the reservoirs of gas or oil have been depleted. Saline aquifers, which are underground pockets of saltwater, are another storage possibility. So are coal seams (where coal can be found) that are so deep underground that they cannot be mined.

While the amount of space available to store the carbon dioxide is limited, there is enough space to be able to store the gas produced for many years. However, there is some danger to storing the carbon dioxide. If it mixes with a fresh-water aquifer (underground stream) or gets to the surface, it could change the chemistry of the soil. Even worse, if the carbon dioxide should leave its storage space and end up in a place that is a depression without wind, the gas, which is heavier than air, could start to collect. If enough carbon dioxide collects, it could suffocate animals or people. This tragedy has happened in the past. In 1986 in Cameroon, 1,800 people died after 87 million cubic yards (80 million cubic meters) of carbon dioxide erupted from a volcanic crater.

Another potential problem with steam reforming is that the natural gas needed for the process is available in only a limited supply, like all fossil fuels. Steam reforming produces hydrogen on a large scale, but a method needs to be developed to do steam reforming on a smaller scale so this reaction can take place either on the vehicle or at a filling station that supplies hydrogen.

USING HYDROGEN

The most commonly researched and most developed application of using hydrogen as a fuel source is in conjunction with a hydrogen fuel cell. Fuel cells operate by mixing hydrogen and oxygen to produce water and electricity. The electricity can then be used to provide power to homes, schools, and even businesses or to power cars and other vehicles. Some experts believe that internal combustion engines (ICEs) that are fueled by hydrogen are just as important. Hydrogen could be used as fuel for transportation by creating internal combustion engines for vehicles that run on hydrogen or hydrogen fuel mixtures.

Using hydrogen in fuel cells

A fuel cell works sort of like a battery. In hydrogen fuel cells, the hydrogen is converted to electricity through an electrochemical reaction. A fuel cell does not run out of power as long as its fuel, hydrogen, is present. There are several types of fuel cells. Some use phosphoric acid as an electrolyte (a substance that conducts electricity). Others use molten carbonate as electrolytes.

The most common type of hydrogen fuel cell in use is the proton exchange membrane (PEM) fuel cell. General Electric first invented this fuel cell in the 1960s as a source of electrical power for the Gemini spacecraft. Though they were expensive, these fuel cells were efficient producers of energy.

PEM fuel cells are usually stacked when they are used in vehicles. That means a number of identical fuel cells are put together to provide a significant amount of energy. The more fuel cells that are put together, the more voltage created. The number of fuel cells stacked in each vehicle varies by the amount of power needed.

Hydrogen fuel cell vehicles

While fuel cells were used early in the United States space program, most discussion of hydrogen fuel cells has focused on vehicles such as cars, buses, and vans. Most major car companies around the world are working on fuel cell technology in some form. Each company has produced its own concept cars and is working toward solving the problems related to building such cars on a mass scale. Even a high-end, limited production company like Rolls Royce has researched hydrogen fuel cells for cars. This company is hoping to have a fuel cell–powered hydrogen prototype completed by 2008. Rolls Royce has been working on hydrogen fuel cell research since 1992.

Daimler Chrysler began research on fuel cells in the 1990s. The company's first fuel cell car was introduced in 1994 and called NECAR 1. Many different versions followed, some of which were tested on the road. In 1997 the car company also introduced a fuel cell bus called the NEBUS. This was followed later with the Mercedes-Benz Citaro bus. About thirty of these buses were used on a test basis in cities throughout Europe between 2003 and 2006.

General Motors (GM) has been working on hydrogen fuel cell technology for many years. The company produced its first fuel cell–powered car in 1966. Though this research area was dropped soon after, GM resumed its work on hydrogen fuel cells in the

A Lockheed Martin Atlas IIIB rocket lifts off the foggy launch pad 36B at Cape Canaveral Air Force Station, early February 3, 2005. The Atlas/Centaur upper stage was powered by burning liquid oxygen and liquid hydrogen. © *Thom Rogers/Corbis.*

early 1980s. By the early 2000s GM had about six hundred employees researching fuel cells. The company formed a partnership with Toyota in 1999 to share hydrogen fuel cell research.

Some of GM's experimental vehicles have been used on a limited basis. In 2003 Federal Express agreed to use one of GM's fuel cell vehicles for one year on normal routes to see how it would work. GM has also conducted test runs of one of its hydrogen fuel cell cars, the HydroGen 3. This vehicle contains 200 hydrogen fuel cells and costs

A zero-emission hydrogen fuel cell bus waits at Aldgate bus station on its first day of service in central London, January 14, 2004. The bus emits only water vapor. © *Toby Melville/ Reuters/Corbis.*

about $1 million to build. HydroGen 3s are being used by the federal government in Washington, D.C., on an experimental basis.

Toyota and Honda also have invested in hydrogen fuel cell technologies. Beginning in 1992 Toyota started working on fuel cell hybrid vehicles, coming up with four prototypes. Road testing of one of the company's fuel cell–powered cars began in 2002. These cars were used at the University of California, Irvine, and University of California, Davis.

Honda began its research into this technology in 1989. Its fuel cell vehicles have been tested on roads in the United States since about 1999. One concept car, the Honda FCX, was tested by the city of Los Angeles in 2002. In 2003 this vehicle was certified for commercial use by the Environmental Protection Agency and the California Air Resources Board.

A number of countries are using hydrogen fuel cell–powered buses on an experimental basis. From 1998 to 2000 several hydrogen-powered buses were used in Chicago and in Vancouver, British Columbia, Canada. British Columbia later bought three other buses to use experimentally in the early 2000s. Vancouver had more buses delivered in 2005 for a further three-year experimental run. In London, England, three of these buses began running in 2003.

Fuel cells as generators

Though most of the media attention has focused on hydrogen fuel cells in vehicles, hydrogen fuel cell–powered generators are already being used in at least 600 buildings around the world. Hospitals, data centers, and office buildings use this technology in their backup generators. Some businesses use these fuel cell generators as part of their source of power. For example, fuel cells provided about 15 percent of the power at a major office building, 4 Times Square, in New York City in 2003.

Using hydrogen in ICEs

When discussing hydrogen as a fuel source, most of the focus in the twentieth and early twenty-first centuries has been on fuel cells. However, some experts believe that internal combustion engines (ICEs) that are fueled by hydrogen are just as important. One early believer in this vision was German researcher Rudolf Erren. He was concerned with the amount of oil his country imported and the emissions that automobiles produced well before most countries took note of these issues. In 1930 he saw that hydrogen could be used as fuel for transportation. He believed that this hydrogen should be produced by water electrolysis. Erren spent time working on creating internal combustion engines for vehicles that could run on hydrogen or fuel mixtures that included hydrogen.

How an Internal Combustion Engine Works

An internal combustion engine (ICE) is a vehicle engine in which the combustion of the fuel takes place within internal cylinders. Virtually all cars today use internal combustion engines, with gasoline as the fuel. A hydrogen ICE is not unlike a gasoline-powered ICE. The hydrogen provides power to create the explosions in the engine that power the car. Inside the engine, pistons move up and down within their cylinders. As each piston pushes up, it compresses a mixture of fuel (hydrogen or gasoline) and air. As the piston reaches the top, the combination of fuel and air is ignited by a spark plug. This explosion forces the piston down inside the cylinder. The ignited fuel also turns the crankshaft in the engine, which eventually leads to the wheels of the car turning. The piston again pushes up in the cylinder to make the exhaust from the ignition move out of the valves located at the cylinder's top. After this step, the piston returns to the bottom of its cylinder. This movement allows another mix of air and fuel to fill the cylinder. This mixture comes in through another set of valves. Then the process begins again.

Hydrogen-powered ICEs are intended for use in buses, cars, vans, and other types of vehicles. Although car manufacturers have already created some hydrogen ICEs, there has not been as much focus on the development of hydrogen ICEs as on hydrogen fuel cells. BMW is one manufacturer that has focused primarily on developing a hydrogen ICE. The company began this research in 1978. Since then BMW has developed several kinds of hydrogen ICEs, which use various hydrogen-to-air ratios, depending on the power desired. The company has also explored using liquid hydrogen as opposed to hydrogen's gaseous form. When liquid hydrogen is used, the car does not need to be refueled as often.

Interestingly, most of BMW's hydrogen ICEs can run on gasoline as well as hydrogen. One BMW concept car that can run on either hydrogen or gasoline is called the H2R. This car was introduced in 2005. The engine in this vehicle is very similar to a standard gasoline ICE that BMW uses in another car, the 760i. Though the engine in the H2R can run on hydrogen, it has an efficiency level similar to a traditional engine. Because the engine in the H2R can run on gasoline or hydrogen, the driver has flexibility in fueling. This quality can be especially important if the hydrogen runs out. A tank of hydrogen only lasts about 215 miles on the H2R, much less than a similar tank full of gas. BMW hopes to sell cars using this type of ICE in Europe by 2007 or 2008. The company wants to put them on the market in the United States by about 2010.

Another car company, Ford, has divided its research focus between hydrogen ICEs and fuel cell cars. The company has developed several hydrogen ICE concept cars, including one car called the Model U and a version of the Ford Focus. Ford also has worked on other vehicles that use hydrogen ICEs, including vans and buses. Ford hopes to have 100 such vans in service by 2006. As for its buses, they were first tested at the 2005 Detroit Auto Show, where they were used as shuttles for reporters. In 2006 the company will sell some of these buses to the state of Florida.

Benefits and drawbacks of existing hydrogen technologies

Each use of hydrogen as fuel has specific benefits and drawbacks. Hydrogen fuel cells are already in use as electrical generators, and they have also been used in the space program. Most experts believe the fuel cell is likely to be the dominant hydrogen technology in the future, not only for electrical generation but also to power vehicles. The only by-product of using a hydrogen fuel cell to power a car is water or water vapor, which exits through the tailpipe. However, hydrogen ICEs are so similar to existing gasoline ICEs that they could be the best first use of hydrogen as a transportation technology for the general public. Also, like fuel cells, hydrogen ICEs do not produce harmful by-products.

Benefits and drawbacks of hydrogen fuel cells

Hydrogen fuel cells have many good aspects. Fuel cells are very easy to make. They contain no moving parts. This means that there is little maintenance that needs to be performed on each fuel cell. Because they have no moving parts, fuel cells are quiet. Fuel cells are also light and versatile. They can be manufactured big or small and used on a large or small scale. Because they are modular in design, one can work on its own or many can function together as one.

Hydrogen fuel cell-powered cars are very efficient producers of power. They are more efficient than internal combustion engine cars. About 60 percent of the potential energy in hydrogen is made into electricity by a fuel cell. These fuel cell-cars can respond instantaneously to provide fuel when it is needed.

Yet there are several major drawbacks to the development and use of fuel cells. One is the lack of a worldwide standard for fuel cells between manufacturers or most governments. Only one standardization agreement was in place as of 2005. It was between Japan and the European Union. This agreement covered hydrogen fuel cells for automobiles. Because no standards are yet in place, the development of the infrastructure needed to support hydrogen technology has been delayed. Governments and businesses do not want to invest money in creating an infrastructure that could be useless if it does not match the standards that others use.

The cost of the energy produced by a fuel cell is also very high. It costs more per kilowatt produced when compared to a gasoline-powered combustion engine. In 2002 a fuel cell could cost anywhere from $500 to $2,500 per kilowatt produced, while the combustion engine only cost about $30 to $35 for the same amount of energy. The costs for fuel cells have been going down as technology has been developed and improved.

Benefits and drawbacks of hydrogen-powered ICEs

One positive aspect to hydrogen-powered ICEs is that engineers at car companies are already experienced in the construction of such engines. The engines are similar to gasoline-powered ICEs. These types of ICEs are more familiar to automotive engineers than the technology of fuel cell engines. These vehicles will also be simpler internally than gasoline-powered cars. The catalytic converters and related systems found on gasoline-powered ICEs to clean up the by-products of fossil fuel combustion are not needed if hydrogen is used.

But hydrogen-powered ICEs have several disadvantages. The cars that use this type of engine are not as efficient as fuel cell-powered cars. Hydrogen ICEs can only extract about half of the chemical energy that is contained in a unit of hydrogen as compared to a fuel cell-powered vehicle. The vehicles also need more space to store fuel than gasoline-powered ICEs. These vehicles are built on current fuel tank sizes designed for gasoline or diesel fuel. Because hydrogen is not a very dense gas, the tanks cannot hold very much hydrogen. Therefore, the vehicles cannot travel as far.

TRANSPORTING HYDROGEN

The form of hydrogen transportation depends on the form of hydrogen being transported. There are different methods for transporting gaseous hydrogen and liquid hydrogen. Most of these

BMW's hydrogen-powered H2R Record Car was styled at its California Designworks USA studio and is powered by a hydrogen-fueled internal combustion engine. © *Ted Soqui/Corbis.*

methods are still being developed and refined; they are not yet in large-scale use.

Transporting gaseous hydrogen

In its gaseous form, hydrogen could be transported over a network of pipelines. Pipelines are commonly used today to distribute hydrogen over a short distance for industrial use, but a wider system would have to be introduced if hydrogen becomes the fuel source of choice for vehicles, homes, and businesses. This pipeline system could be similar to the way that natural gas is distributed. The hydrogen pipeline system also would need more compressors than a natural gas system. A small amount of hydrogen that is traveling along the pipeline would have to be used to power the compressors. Some experts believe that one way to address the distribution question is by converting natural gas pipeline systems to hydrogen. These supporters believe that only the seals, the meters, and the equipment at the end of the pipeline would have to be modified to support hydrogen. There are also trucks that transport hydrogen as a compressed gas, but they hold a much smaller quantity than a gasoline tanker.

Transporting liquid hydrogen

Transporting the liquid form of hydrogen could take many forms. As gasoline is now, hydrogen could be transported via truck, railcar, or ship. This method could be expensive and difficult. It would take about 21 tanker trucks of hydrogen to carry the equivalent of one gasoline tanker because hydrogen has a low density.

Benefits and drawbacks of hydrogen transport methods

The infrastructure to transport hydrogen does not yet exist. Some experts believe that the questions about how to produce, distribute, and store the hydrogen have to be answered all at once for the infrastructure to be properly implemented. Regardless of which methods are eventually used, it will still cost billions of dollars to create this transportation infrastructure. That cost is one large obstacle to the development of better transportation methods.

DISTRIBUTING HYDROGEN

At least in the case of hydrogen-powered vehicles, the primary means by which hydrogen would be distributed for public consumption is through a hydrogen filling station. Such a station would be like a gas station, only with hydrogen instead of gasoline. As of 2005 there were only about 100 hydrogen filling stations in existence in the world.

By 2005 the Clean Urban Transport for Europe program was expected to build several hydrogen filling stations in major European cities. Germany is especially committed to building hydrogen filling stations. The German government is helping to pay for the building of the self-sufficient hydrogen filling stations as a step toward the hydrogen economy.

The United States government has also made a commitment to building hydrogen filling stations. In 2004 the U.S. Department of Energy promised to spend $190 million to build gas stations that would offer both hydrogen and gasoline. The money is also intended to support other projects related to the development of the infrastructure needed to support the hydrogen economy. This money will be spent, however, only if private industry will match the amount.

A few hydrogen filling stations already exist in the United States. In 2005 in Washington, D.C., the first hydrogen-gasoline fueling station was opened by Shell. It provides hydrogen for the six fuel-cell cars that General Motors provided to the area. Both the cars and the station were demonstrations to show the potential of hydrogen as a fuel source. The state of California is also committed to building hydrogen filling stations. By 2010 the California government has promised to have 150 to 200 hydrogen fueling stations on the interstate highways in California as part of the California Hydrogen Highway Network. They will be located on all 21 of the state's interstate freeways. Under the California plan, hydrogen filling stations will be found every 20 miles to provide convenient access for consumers.

Benefits and drawbacks of hydrogen distribution methods

One large benefit to using filling stations to distribute hydrogen fuel is that consumers all over the world already use such stations to fill their gasoline-powered cars. The general public would not need to be educated on the concept of using filling stations for their automobiles.

However, there are drawbacks with this technology. In Europe, for example, the electrolysis system is often employed to convert water to hydrogen at the filling stations. The problem with this kind of filling station is the large amount of electricity needed to make the conversion possible. Electricity is expensive, and current electricity generation depends heavily on fossil fuels. In Germany, experiments are being conducted to use wind as a source of electricity for on-site electrolysis at filling stations. In the United States, wind-driven on-site electrolysis at filling stations is not seen as feasible in most parts of the country. Instead, biomass is the method being examined. In this process, waste from logging and lumber as well as leftover crop plants is used to produce the electricity needed.

In addition to working on the technology behind hydrogen filling stations, governments and companies have to build the stations. The cost will be enormous, and many governments have pledged funds for this to happen.

STORING HYDROGEN

Hydrogen is usually stored as a liquid, though it can also be stored as a gas or a solid. Because hydrogen is low in density, storing it is a challenge. This is true both for storage at hydrogen production sites as well as on vehicles that might use hydrogen as a fuel. Among the methods for storing hydrogen are the following:

- Compressing it into cylinders of various sizes. This is one of the most common ways to store hydrogen for industrial use.
- Using compressed gas tanks for vehicles. Many automotive manufacturers and researchers have been experimenting with these tanks. Instead of cylinders, hydrogen would be pumped into a compressed gas tank on the car and stored there.
- Storing liquid hydrogen cryogenically (at very low temperatures).

Benefits and drawbacks of storage options

Storage of hydrogen on vehicles is a major concern. Some scientists believe that the storage of hydrogen on cars is the biggest single problem facing the use of hydrogen as a fuel for cars. Vehicles have very limited space for storing hydrogen, and the amount that needs to be stored for hydrogen to be a viable fuel source is rather large.

As mentioned, hydrogen is usually stored as a liquid. However, liquid hydrogen has many drawbacks. For example, liquid hydrogen has to be stored at temperatures at or below −423°F (−253°C). To keep the liquid this cold requires a significant amount of energy. The system also must be insulated. Also, even if liquid hydrogen is stored at the right temperature, about three to four percent is boiled off daily. This situation could be a problem for vehicles that are not being used for a few days at a time.

Because of the low density of hydrogen, the amount of hydrogen that can be compressed into a cylinder is less than more dense substances. This problem means that compression has a significant energy cost and an economic expense. The cylinders also must be transported from the place the hydrogen is manufactured to the market where it is needed.

The same drawback hinders compressed gas tanks on vehicles. As of 2005 most compressed gas tank systems can only carry about 5,000 pounds per square inch (psi) of hydrogen. For the ideal range for a car, researchers hope to develop a tank system that offers 10,000 psi. For now compressed gas tanks are large and hard to fit onto a car. They are also made from materials that are both heavy and expensive. One such material is carbon fiber. There are also safety concerns for hydrogen compressed gas tanks. To be safe, they must be able to withstand a very powerful impact. This is a goal that has not been fully reached in a workable manner.

IMPACTS

Using hydrogen as an alternative energy source would have numerous impacts. Perhaps the biggest would be in the environmental arena, as the development of hydrogen-powered vehicles could drastically reduce the pollution that contributes to global warming, depending on the production method. In addition, because the fossil fuels that currently are used for most of the world's power will one day run out, society will need to find alternative energy sources to power its homes, businesses, and transportation needs. Hydrogen can be an important part of this alternative future. However, not all of the potential impacts are positive ones.

Environmental impact

Much of the impact of adopting hydrogen as an energy source would be positive for the environment. The use of hydrogen would likely come with a reduction of the use of fossil fuels as energy sources. With this reduction would perhaps come a reduction in global warming, because fossil fuel use is believed to be an important contributor to global warming.

However, the production of hydrogen can potentially affect the environment in a negative way. Depending on the production method, carbon dioxide and other negative emissions can enter the atmosphere while hydrogen is being made. This issue can be addressed by catching and storing the carbon dioxide, but even this storage can potentially affect the environment. However, if environmentally friendly, renewable resources such as solar or wind are used to power the means of producing hydrogen, the negative impact can be eliminated.

Another potential problem is that if hydrogen becomes widely used, it could leak into the atmosphere. If the amount is significant enough, this hydrogen could change the percentage of hydrogen present in Earth's atmosphere. Some scientists believe that this could have a profound effect on the atmosphere, including increasing the size of the hole in the ozone layer. More hydrogen in the atmosphere could also lead to more high altitude clouds and increase the number of soil microbes that rely on hydrogen as their primary nutrient. The soil microbe increase could change the ecology of Earth. However, there are soil micro-organisms that consume hydrogen as well, and they might be able to balance these problems out. The outcome of putting more hydrogen in the atmosphere is uncertain.

A final environmental question is what to do with the water or water vapor that would be produced by cars using hydrogen fuel cells. Since such water is pure, it will freeze in temperatures below 32°F (0°C). Scientists will have to come up with a solution for this by-product on the roadways and the environment in colder climates.

Economic impact

Adopting a hydrogen-based economy could lead to an extreme change in a number of industries. The way the automotive business would be run would change completely as these companies focused on building cars, trucks, and buses that use hydrogen instead of gasoline. The oil/petroleum business would suffer at some point as the use of hydrogen creates less dependence on oil. The adoption of hydrogen could also impact the electric industry, especially if electrolysis is widely adopted as a means of producing hydrogen.

Whole new industries would also be created as the infrastructure needed to support hydrogen is put in place. The production, transportation, distribution, and storage of hydrogen could have a huge economic impact as billions of dollars would be invested around the world to create the infrastructure for the hydrogen economy. As this infrastructure is put in place, those who could fix and maintain hydrogen filling stations, production plants, generators, vehicles, and other such hardware would be needed. This would create new jobs and businesses.

Automotive manufacturers in 2005 expect hydrogen-powered ICE cars to hit the marketplace within five to ten years. Because the public might embrace hydrogen-powered ICEs more easily than fuel cell-powered cars, some observers believe that if these kinds of vehicles can get on the market, the hydrogen economy can grow rapidly. The spread of cars with hydrogen ICEs would create a demand for hydrogen fuel and a place to buy it.

The development of hydrogen fuel cells would also have an economic impact. In addition to creating an industry for the production of fuel cells themselves, the manufacturing processes used for vehicles, generators, and other products that use fuel cells would change.

Societal impact

The implementation of the hydrogen economy would affect society worldwide. In countries that are already developed, such as the United States and Great Britain, sources of power and the way vehicles run and even sound would be different. Fueling cars would also be a somewhat different experience than it is right now.

Hydrogen could also change the way the whole power grid works. Currently, developed countries receive their power from centralized power stations. These stations produce the electrical power from fossil fuels of one kind or another and then send the power through wires to individual businesses and homes. If a power station goes out, all the homes and businesses connected to it on the grid also go out. In a hydrogen-based system, individual fuel cell sites could generate electricity for homes and businesses independently. If the overall power grid were to become less centralized, it would be less vulnerable to terrorist attacks aimed at crippling a nation's energy supply.

Even hydrogen fuel cell–powered vehicles might act as small generators and provide power for others when they are not in use. The cars would be plugged into something like wall sockets. The fuel cells on the cars could power the local electrical power grid, instead of the grid providing electricity. According to one estimate, only 4 percent of hydrogen fuel cell-powered cars working in this fashion could provide enough power for an entire city.

The impact of major hydrogen use would be even greater on countries that were underdeveloped or undeveloped. Especially if hydrogen is made with a renewable fuel resource such as solar or wind power, energy could be easily accessible to every country on Earth. Developing countries would have better, easier access to electricity and other forms of energy. They could make their own hydrogen energy rather than importing oil to use in generating electricity. The hydrogen economy could better the lives and economies of everyone as local industries spring up, jobs are created, and opportunities abound for social and economic improvement.

In addition to making the United States and other countries less dependent on nonrenewable sources of energy such as oil, hydrogen fuel cell-powered cars in particular could affect noise pollution. Because fuel cell-powered vehicles are very quiet, the familiar sounds of gasoline-powered internal combustion engines would be gone. Urban noise pollution in particular would be greatly lessened, providing a more peaceful environment.

On the other hand, there are a number of safety issues related to the implementation of hydrogen. One problem is that when hydrogen burns, the flame is invisible. In other words, the fire produced by hydrogen is hard to see. The gas itself can also leak out without being detected. Any build up of gas could lead to dangerous explosions, because, although hydrogen is very light weight, is diffuses rapidly. These issues have to be addressed. The first problem could be solved by adding something to the gas so it burns in a way that people can see. One way to solve the second problem is by creating warning instruments that can detect hydrogen gas leaks in the container or the supply chain. Also, colorants can be added to the hydrogen so that the leaks are more easily noticed.

FUTURE TECHNOLOGY

The future of hydrogen as a fuel source might include power plants based on hydrogen technology. Other means of transportation might also benefit from the use of hydrogen as a fuel. For example, planes could take advantage of the fact that hydrogen weighs less than conventional fuels.

Some researchers believe that hydrogen fuel cell-powered generators will be implemented before cars using that technology become widespread. In a 2004 article in *Scientific American,* Matthew L. Wald noted, "Although most people may have heard of fuel cells as alternative power sources for cars, cars may be the last place they'll end up on a commercial scale." Instead, Wald and others believe that consumer products such as laptop computers, video cameras, and cell phones could be among the first items to be powered by hydrogen fuel cells. Fuel cells are also expected to provide electricity for homes and businesses. Hydrogen fuel cells could potentially provide a source of electric power for electric utilities and in power plants.

For hydrogen fuel cells to become a cornerstone of the hydrogen economy, technological advances must make them cheaper to produce and more powerful when in operation. For example, scientists are working on ways to lessen the need for the platinum catalysts used in PEM fuel cells. Platinum is an expensive precious metal that can add to the cost of building a fuel cell.

CONCLUSION

There are many technological and economic hurdles to adopting hydrogen as an alternative energy source. Still, many experts believe that hydrogen will be the primary energy source of the twenty-first century and beyond. Perhaps more than any other alternative technology that currently exists, hydrogen has the potential to replace our dependence on fossil fuels with a clean source of energy that will never run out.

For More Information

Books

Ewing, Rex. *Hydrogen: Hot Cool Science—Journey to a World of the Hydrogen Energy and Fuel Cells at the Wassterstoff Farm.* Masonville, CO: Pixyjack Press, 2004.

Rifkin, Jeremy. *The Hydrogen Economy.* New York: Tarcher/Putnam, 2002.

Romm, Joseph J. *The Hype of Hydrogen: Fact and Fiction in the Race to Save the Climate.* Washington, DC: Island Press, 2004.

Periodicals

Behar, Michael. "Warning: The Hydrogen Economy May Be More Distant Than It Appears." *Popular Mechanics* (January 1, 2005): 64.

Burns, Lawrence C., J. Byron McCormick, and Christopher E. Borroni-Bird. "Vehicles of Change." *Scientific American* (October 2002): 64-73.

Graber, Cynthia. "Building the Hydrogen Boom." *OnEarth* (Spring 2005): 6.

Grant, Paul. "Hydrogen Lifts Off—with a Heavy Load." *Nature* (July 10, 2003): 129-130.

Guteral, Fred, and Andrew Romano. "Power People." *Newsweek* (September 20, 2004): 32.

Hakim, Danny. "George Jetson, Meet the Sequel." *New York Times* (January 9, 2005): section 3, p. 1.

Lemley, Brad. "Lovin' Hydrogen." *Discover* (November 2001): 53-57, 86.

Lizza, Ryan. "The Nation: The Hydrogen Economy; A Green Car That the Energy Industry Loves." *New York Times* (February 2, 2003): section 4, p. 3.

McAlister, Roy. "Tapping Energy from Solar Hydrogen." *World and I* (February 1999): 164.

Muller, Joann, and Jonathan Fahey. "Hydrogen Man." *Forbes* (December 27, 2004): 46.

Port, Otis. "Hydrogen Cars Are Almost Here, but . . . There Are Still Serious Problems to Solve, Such As: Where Will Drivers Fuel Up?" *Business Week* (January 24, 2005): 56.

Service, Robert F. "The Hydrogen Backlash." *Science* (August 13, 2004): 958-961.

Terrell, Kenneth. "Running on Fumes." *U.S. News & World Report* (April 29, 2002): 58.

Wald, Matthew L. "Questions about a Hydrogen Economy." *Scientific American* (May 2004): 66.

Westrup, Hugh. "Cool Fuel: Will Hydrogen Cure the Country's Addiction to Fossil Fuels?" *Current Science* (November 7, 2003): 10.

Westrup, Hugh. "What a Gas!" *Current Science* (April 6, 2001): 10.

Web Sites

''Driving for the Future.'' California Fuel Cell Partnership. http://www.cafcp.org (accessed on August 8, 2005).

''How the BMW H2R Works.'' How Stuff Works. http://auto.howstuffworks.com/bmw-h2r.htm (accessed on August 8, 2005).

''Hydrogen, Fuel Cells & Infrastructure Technologies Program.'' U.S. Department of Energy Energy Efficiency and Renewable Energy. http://www.eere.energy.gov/hydrogenandfuelcells/ (accessed on August 8, 2005).

''Hydrogen Internal Combustion.'' Ford Motor Company. http://www.ford.com/en/innovation/engineFuelTechnology/hydrogenInternalCombustion.htm (accessed on August 8, 2005).

''Reinventing the Automobile with Fuel Cell Technology.'' General Motors Company. http://www.gm.com/company/gmability/adv_tech/400_fcv/ (accessed on August 8, 2005).

CD-ROMs

World Spaceflight News. *21st Century Complete Guide to Hydrogen Power Energy and Fuel Cell Cars: FreedomCAR Plans, Automotive Technology for Hydrogen Fuel Cells, Hydrogen Production, Storage, Safety Standards, Energy Department, DOD, and NASA Research.* Progressive Management, 2003.

Nuclear Energy

INTRODUCTION: WHAT IS NUCLEAR ENERGY?

Nuclear energy is energy that can be released from the nucleus of an atom. There are two ways to produce this energy, either by fission or fusion. Fission occurs when the atomic nucleus is split apart. Fusion is the result of combining two or more light nuclei into one heavier nucleus. Most often, when people discuss nuclear power, they are talking about nuclear fission. Power production from fusion is still in its infancy.

Atoms are made up of several parts: protons, neutrons, electrons, and a nucleus. A nucleus is the positively charged center of an atom. Protons are positively charged particles, and neutrons are uncharged particles. Electrons orbit around the nucleus and are negatively charged. Fission can occur in two ways — first, in some very heavy elements, such as rutherfordium, the nucleus of an atom can split apart into smaller pieces spontaneously. With lighter elements, it is possible to hit the nucleus with a free neutron, which will also cause the nucleus to break apart.

Either way, a significant amount of energy is released when the nucleus splits. The energy released takes two forms: light energy and heat energy. Radioactivity is also produced. Atomic bombs let this energy out all at once, creating an explosion. Nuclear reactors let this energy out slowly in a continuous chain reaction to make electricity. After the nucleus splits, new lighter atoms are formed. More free neutrons are thrown off that can split other atoms, continuing to produce nuclear energy. The first controlled nuclear reaction took place in 1942.

Nuclear fission

Since at least the 1920s, scientists had believed that it might someday be possible to produce energy by splitting atoms. They

Words to Know

Critical mass An amount of fissile material needed to produce an ongoing nuclear chain reaction.

Decay The breakdown of a radioactive substance over time as its atoms spontaneously give off neutrons.

Enrichment The process of increasing the purity of a radioactive element such as uranium to make it suitable as nuclear fuel.

Fission Splitting of an atom.

Fusion The joining of atoms to produce energy.

Meltdown Term used to refer to the possibility that a nuclear reactor could become so overheated that it would melt into the earth below.

Pile A mass of radioactive material in a nuclear reactor.

based this belief on their growing understanding of the physics of the atom. They knew that atoms contain energy, and they believed that by "splitting" the atom, or breaking it apart, they could release that energy. The process would come to be called nuclear fission.

An atom is made up of three kinds of particles: neutrons, protons, and electrons. Two of these particles, neutrons and protons, are found in the nucleus, or center, of an atom. A neutron does not have an electrical charge. It is called a neutron because its electrical charge is neutral. A proton has a positive electrical charge. Circling around the nucleus of an atom in layers are electrons, which have a negative electrical charge. To keep the overall electrical charge neutral, an atom has to have the same number of protons and electrons. Positive and negative electrical charges attract each other. The charges bind the particles of an atom together. When an atom is split, some of this energy is released.

The atoms of different elements have different numbers of particles. Some elements are very simple and light. Hydrogen is the simplest and lightest element because it has only one proton, one electron, and no neutrons. In contrast, the heaviest element in nature is uranium. (Some heavier elements have been artificially produced in laboratories, but these elements do not exist in nature.) Uranium atoms contain ninety-two protons and ninety-two electrons. The number of neutrons can vary, depending on the isotope of uranium under consideration. An isotope is a "species" of an element. It contains a different number of neutrons from other isotopes of the same element. Generally, uranium nuclei contain either 143 or 146 neutrons.

Spontaneous Fission

Some elements, including uranium, undergo fission spontaneously, or on their own, as neutrons break away from the atom. These elements are said to be radioactive because they release subatomic particles and energy. This spontaneous fission is generally a very slow process. Scientists use the word *decay* to refer to the breakdown of a radioactive substance over time as it releases its neutrons.

For nuclear energy, uranium is the most important element. Uranium is used as fuel to produce nuclear reactions. It makes a good fuel source because uranium atoms are so big and heavy. They are easier to break apart. These large atoms can be thought of as a house built with playing cards. The house becomes increasingly unstable as cards are added, and is more likely to fall apart the bigger and heavier it gets. In a nuclear power plant, the goal is to create fission from uranium fuel and to be able to speed the reaction up (or slow it down) to control the amount of energy being produced.

HISTORICAL OVERVIEW: NOTABLE DISCOVERIES AND THE PEOPLE WHO MADE THEM

Scientists such as Enrico Fermi (1901–1954) noticed that the free neutrons in elements such as uranium bombard other uranium atoms. This bombardment causes the other atoms to split and release additional neutrons. These additional neutrons then bombard other atoms. The process continues in a chain reaction, or a reaction that keeps going on its own. A neutron in this way can be thought of as similar to a cue ball on a pool table. The cue ball bombards the cluster of balls at the other end of the table, causing the cluster to break apart. All the balls then bounce around, bumping into one another, causing further collisions, and so on.

Fermi had conducted experiments in nuclear fission in 1934 while he was still living in Rome, Italy. He had bombarded uranium with neutrons and discovered that what was left over afterwards were elements that were much lighter than uranium. This led him to believe that the uranium atoms had been split. The mass number of the leftover elements was smaller, so the uranium must have transformed into different elements as it broke down. In 1938

The Periodic Table of the Elements

Elements are the fundamental building blocks of nature. Each box in the Periodic Table of the Elements provides basic information about the size and weight of each element. It arranges the elements from lightest to heaviest. It also arranges them into families that share some important characteristics. Each element has a name and a chemical symbol. In the case of uranium, the symbol is simple, U. The symbols for some elements seem strange. The symbol for lead, for example, is Pb because the symbol is taken from the Latin word for lead, *plumbum.*

The periodic table also contains each element's atomic number and atomic weight. The atomic number is found in the upper left-hand corner of the element's box. It specifies the number of protons in the element's nucleus. Thus, it is equal to the number of electrons. The atomic number for hydrogen is 1, for uranium, 92.

At the bottom center of each box is the element's atomic weight. Atomic weight is a little more complicated. Basically, it represents the combined total of protons and neutrons in the nucleus, called the mass number. But the atomic weight of uranium is given as 238.02891 rather than just 238. The reason for the digits to the right of the decimal point is that many elements, including uranium, occur in different isotopes.

Uranium, for example, has sixteen different isotopes, though only three are found with any frequency. These isotopes are U_{234}, U_{235}, and U_{238}. (Sometimes scientists write these differently, as ^{234}U and so on or U-234.) While the number of protons and electrons in a given element is always the same, the number of neutrons can vary, producing different isotopes. This accounts for the different atomic weights (234, 235, and 238 for uranium). For uranium and other elements, the odd digits to the right of the decimal point occur because on the Periodic Table scientists provide a weighted average of the different isotopes. Therefore, the number may not be a whole number. As a practical matter, the atomic weight figure can be rounded off to the closest whole number.

German scientists Otto Hahn (1898–1968) and Fritz Strassman (1902–1980) conducted a similar experiment. They discovered that what was left over after bombarding uranium with neutrons

The Italian Navigator

In December 1942 a message was sent to a number of high officials in the U.S. government. The message was written in code because at the time, the United States was at war and the authorities wanted to keep the contents of the message secret. The message read: "The Italian navigator has just landed in the new world."

The "Italian navigator" was physicist Enrico Fermi. Fermi had left his native Italy for the United States in 1938 because he saw the storm clouds of World War II (1939–1945) gathering over Europe. "The new world" referred to the successful outcome of an experiment. The experiment was conducted by Fermi and a team of researchers at the University of Chicago. On December 2, 1942, in a squash court under the athletic stadium, Fermi oversaw the world's first controlled nuclear reaction. On that date, humanity did indeed land in a new world, the world of nuclear energy.

was the much lighter element barium. This experiment confirmed that the uranium atoms had split.

Other scientists such as Lise Meitner (1878–1968) from Austria and Niels Bohr (1885–1962) from Denmark arrived at similar results. But they also made a startling discovery. When the atomic weights of the by-products of their experiment were added together, something was missing. If every piece of a broken window is swept up and weighed, the total weight of the pieces should be the same as the weight of the original window. Scientists expected that the same principle would apply to atoms. If atoms broke down because of fission, the atomic weight of the new elements formed, when added together, should be the same as the atomic weight of the original uranium. But Meitner and Bohr found that the elements in the reaction lost mass. Some of the mass had changed to energy. In this way they proved the truth of the famous equation from Albert Einstein (1879–1955), $E = mc^2$. This equation says that energy (E) is equal to mass (m) multiplied by the speed of light (c) squared. Mass, or matter, could be converted into energy.

None of these experiments produced a chain reaction, or a continuing fissioning of atoms. However, in 1942 Fermi thought of a way to create such a chain reaction. He took 40 tons of

Lise Meitner

Lise Meitner's contributions are often overlooked in the history of nuclear power development. As a woman, Meitner was barred from higher education in her native Austria until 1901, when she began studying physics at the University of Vienna. After she completed her doctorate in 1907, she worked with the famous German physicist Max Planck (1858–1947) and chemist Otto Hahn.

Meitner was born into a Jewish family. Although she had converted to Christianity, she was still driven out of Austria and Germany after the Nazi regime took power. She settled in Stockholm, Sweden, where she continued her work on radioactivity. There she worked with Hahn and Strassman. She and another physicist, Otto Frisch (1904–1979), actually coined the phrase ''nuclear fission.''

One of science's worst scandals took place in 1945. That year, Otto Hahn was given the Nobel Prize in Chemistry for the discovery of nuclear fission. The contributions of Lise Meitner were entirely ignored. While such names as Planck, Fermi, Hahn, Einstein, and others were famous in the scientific community, Meitner's name was largely forgotten. Later scientists acknowledged her important role, and in 1966 she was awarded the U.S. Fermi Prize in Physics.

uranium, a nuclear "pile," and surrounded it with 385 tons of graphite blocks to contain the uranium. (A "pile" of nuclear materials is not literally a pile. "Pile" refers to a quantity of nuclear materials in a nuclear reactor.) This would provide him with the "critical mass" needed to produce an ongoing atomic reaction.

Fermi's main concern was to make sure that the reaction did not get out of control. A controlled chain reaction produces a flow of energy, but an uncontrolled chain reaction produces an explosion. Fermi needed a way to make sure that he did not blow up Chicago by letting his planned reaction get out of control. The graphite blocks would help, but he also inserted rods made of cadmium, a soft bluish-white element, into the pile. Cadmium absorbs neutrons, so it can keep nuclear fission reactions under control.

On that December afternoon in 1942, Fermi and his team slowly pulled a few of the cadmium rods out of the pile. Now some of the

A spent nuclear fuel rod in a cooling pond glows a bright blue. Once the rods are used up, they are hot and radioactive. Water-filled pools are sometimes used to cool and store the fuel rods. *© Tim Wright/Corbis.*

spontaneously released neutrons in the uranium could bombard other uranium atoms. Each collision produced an average of 2.5 new free neutrons, which in turn bombarded other atoms, releasing 2.5 more free neutrons, and so on. More rods were slowly pulled out, and the pace of the reaction increased. When rods were pushed back in, the reaction slowed as the cadmium soaked up

The World's First Nuclear Reactor

Enrico Fermi is credited with building the world's first nuclear reactor. Strictly speaking, this is only partially true. He actually built the first "artificial" nuclear reactor. In 1972 a team of French scientists came across an old mine in West Africa. Inside they found some uranium ore. In this ore they found concentrations of U_{235} of 0.4 percent. But the concentration of U_{235} in uranium ore found in nature is always 0.72 percent. By analyzing the trace elements in the ore, the scientists concluded that the amount of U_{235} was less than normal because a chain reaction had occurred. In other words, a naturally occurring nuclear reactor had developed in the mine. The scientists estimate that the reaction occurred more than two billion years ago over a period lasting about 600,000 to 800,000 years.

neutrons. Chicago did not blow up, and Fermi had created the world's first nuclear reactor.

From the Manhattan Project to Atoms for Peace

Fermi conducted his successful experiment almost exactly one year after the Japanese attacked the U.S. naval base at Pearl Harbor, Hawaii, on December 7, 1941. This event pulled the United States into World War II. The war had begun in September 1939, when German dictator Adolf Hitler (1889–1945) ordered his troops to invade Poland. In the years that followed, Germany occupied much of Europe. Meanwhile, the Japanese empire was spreading throughout Asia and the Pacific.

Most of the leading scientists involved in nuclear research were from Germany. U.S. policy makers learned that German scientists were trying to develop an atomic bomb, a bomb whose enormous destructive force would come from an uncontrolled fission reaction. Such a bomb in the hands of Germany could have changed the outcome of the war. Thus, American policy makers developed a plan for the United States to create such a bomb first. This is the reason for the secrecy surrounding the message informing the government that Enrico Fermi's experiment had been successful.

The research program to develop the bomb was the Manhattan Project. (The name Manhattan has no particular meaning. The

Albert Einstein
Old Grove Rd.
Nassau Point
Peconic, Long Island

August 2nd, 1939

F.D. Roosevelt,
President of the United States,
White House
Washington, D.C.

Sir:

Some recent work by E.Fermi and L. Szilard, which has been communicated to me in manuscript, leads me to expect that the element uranium may be turned into a new and important source of energy in the immediate future. Certain aspects of the situation which has arisen seem to call for watchfulness and, if necessary, quick action on the part of the Administration. I believe therefore that it is my duty to bring to your attention the following facts and recommendations:

In the course of the last four months it has been made probable - through the work of Joliot in France as well as Fermi and Szilard in America - that it may become possible to set up a nuclear chain reaction in a large mass of uranium,by which vast amounts of power and large quantities of new radium-like elements would be generated. Now it appears almost certain that this could be achieved in the immediate future.

This new phenomenon would also lead to the construction of bombs, and it is conceivable - though much less certain - that extremely powerful bombs of a new type may thus be constructed. A single bomb of this type, carried by boat and exploded in a port, might very well destroy the whole port together with some of the surrounding territory. However, such bombs might very well prove to be too heavy for transportation by air.

First page of a letter dated August 2, 1939 from Albert Einstein to President Roosevelt discussing the possibilities and implications of nuclear research. © *Corbis.*

branch of the army that oversaw the project was based in Manhattan, New York.) Beginning in 1943, the nation's top scientists, many of them from top-ranked universities, came to Los Alamos, New Mexico. The brilliant physicist J. Robert Oppenheimer (1904–1967) directed the research. They worked in shacks and lived in primitive conditions, all the while keeping their work top secret.

Continuing the research of Fermi and others, the scientists succeeded in building an atomic bomb, which they tested in the New Mexico desert on July 16, 1945. By this time, though, Germany had surrendered and the war in Europe was over. The war continued to rage in the Pacific as the United States and its allies fought the determined Japanese empire. During the final months of the war with Japan, both countries lost large numbers of troops in bloody island battles, such as those on the Japanese island of Iwo Jima. The Japanese were defeated, but the nation refused to surrender. To put a quick end to the war, the United States released an atomic bomb over the Japanese city of Hiroshima on August 6, 1945. A similar bomb destroyed Nagasaki three days later. Together, the two bombs immediately killed over one hundred thousand people, and many more would later die as a result of burns and radiation sickness. Faced with such a destructive weapon, the Japanese finally surrendered.

The decision to use the atomic bomb was highly controversial. Many U.S. policy makers urged use of the bomb as a way to save the lives of U.S. (and Japanese) troops, who faced the possibility of a difficult invasion of Japan. Others, including many nuclear scientists, believed that using the bomb would cause too much destruction and death. Many believed that it was just a matter of time before Japan would surrender.

After the Soviet Union developed its own atomic weapons, the world's two superpowers began to stockpile them. They accumulated far more nuclear weapons than would ever be needed to defeat the other side. In the 1950s and beyond, the world lived in fear that a nuclear war would erupt, with devastating consequences. Scientists, though, searched for peaceful ways to use nuclear energy. On December 8, 1953, U.S. president Dwight D. Eisenhower (1890–1969) addressed the United Nations. In his speech, he outlined the "Atoms for Peace" program. He suggested that atomic development and research be turned over to an international agency and that research be conducted to find peaceful uses for atomic energy. This speech gave a major push to efforts to harness atomic energy for the benefit of humankind rather than as a weapon.

Atomic energy development

Those efforts had already begun in the United States. In 1946 the government created the Atomic Energy Commission. Its job was to oversee the development of nuclear power. One of its first steps was to authorize the development of Experimental Breeder Reactor I in Arco, Idaho. On December 20, 1951, the reactor produced the world's first electricity fueled by nuclear power, lighting four 200-watt light bulbs. On July 17, 1955, Arco, home to one thousand people, became the world's first town to be powered by nuclear energy.

Until this time, nuclear energy had been firmly under the control of the military. The first civilian power plant began operating in Susana, California, on July 12, 1957. The world's first commercial-sized nuclear power plant reached full operating power in 1957 in Shippingport, Pennsylvania. (Most nuclear power plants, for safety reasons, operate at about 70 to 90 percent of their maximum capacity.) Meanwhile, on July 14, 1952, the keel had been laid for the world's first nuclear-powered submarine, the *Nautilus*. On March 30, 1953, the sub powered up its nuclear generators for the first time.

Nuclear power developed rapidly in the late 1950s and into the 1960s. On October 15, 1959, the Dresden-I Nuclear Power Station came online (that is, began to operate) in Illinois. This was the first nuclear power plant to be built entirely without money from the government. On August 19, 1960, the Yankee Rowe Nuclear Power Station in Massachusetts became the nation's third nuclear power plant. On November 22, 1961, the U.S. Navy commissioned the U.S.S. *Enterprise,* the world's largest ship. Powered by nuclear energy, the aircraft carrier could operate at speeds up to 30 knots for as far as 400,000 miles (740,800 kilometers) without having to refuel. Another milestone was passed on December 12, 1963, when the Jersey Central Power and Light Company launched construction of the Oyster Creek nuclear power plant. This was the first nuclear plant to be ordered as an economic alternative to a fossil-fuel plant.

By 1971 the United States was operating twenty-two nuclear power plants that provided 2.4 percent of the nation's electricity. By the end of the 1970s, seventy-two plants were producing 12 percent of the nation's electricity. And by the end of the 1980s, 109 power plants were generating 14 percent of the nation's

The world's first nuclear powered submarine, the U.S.S. *Nautilus*. © *Bettmann/ Corbis.*

electricity. These numbers peaked in 1991, when the number of plants rose to 111, together supplying about 22 percent of the nation's electricity. By the early 1990s nuclear power plants were generating more power in the United States than all power sources combined generated in 1956.

Similar developments were taking place worldwide. As of late 2005, 441 nuclear reactors were producing 2,618.6 billion kilowatt-hours of electricity in thirty countries. The United States led the way with 103 nuclear reactors still in operation. Other countries with a large number of nuclear reactors included Canada (18), France (59), Germany (17), Japan (55), Russia (31), and the United Kingdom (23). The country that generated the highest percentage of its electricity needs from nuclear power was France, at 78 percent. Close behind was Lithuania, whose one power plant generated 72 percent of the nation's electricity.

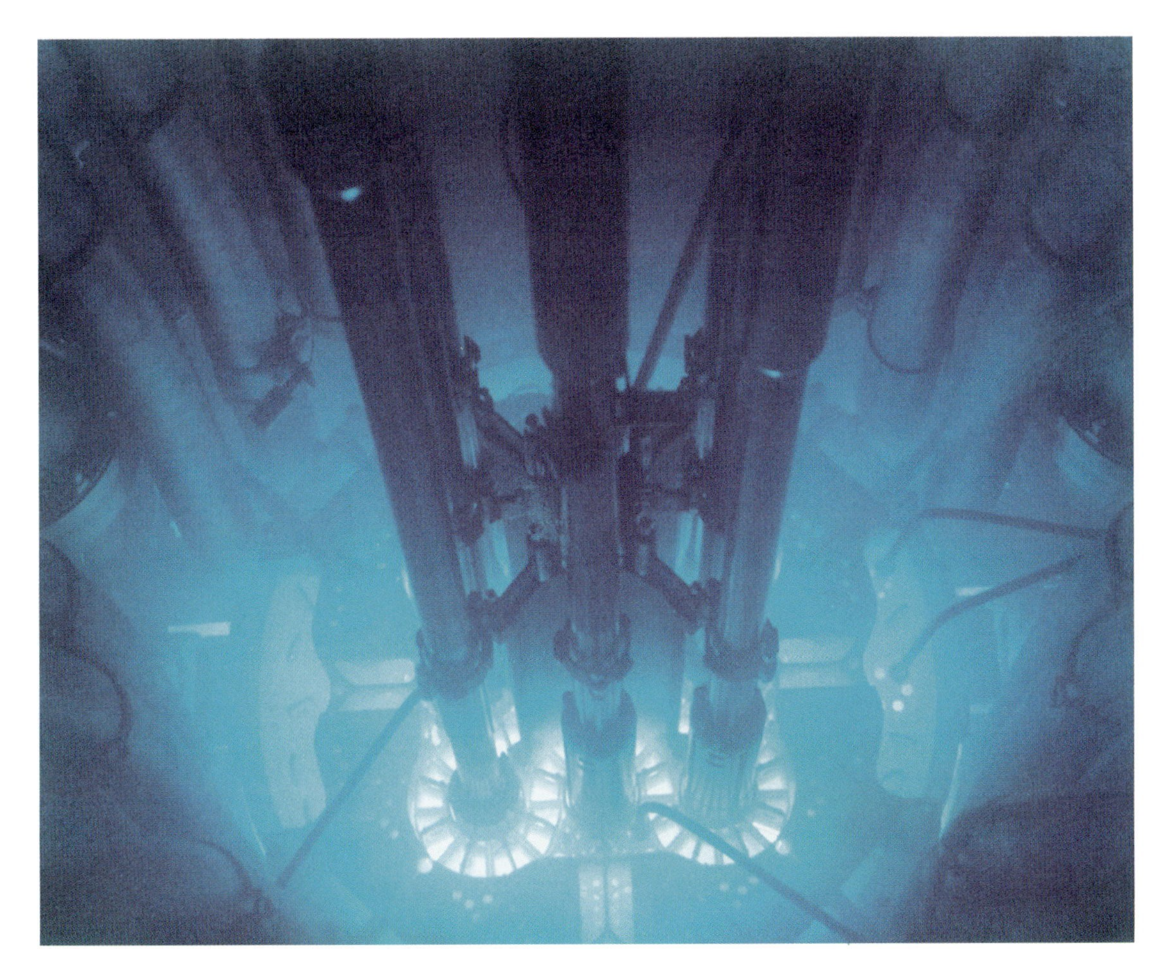

This large reactor in Idaho, USA, operates with a thermal power of 250,000 kilowatts. The reactor is water-cooled and the blue glow results from Cerenkov radiation, emitted when energetic charged particles travel faster through the water than light. *United States Department of Energy/Photo Researchers, Inc.*

Setbacks

In the 1950s and 1960s scientists around the world believed that nuclear power had unlimited potential. Along with most of the public, they believed that nuclear plants would provide an endless source of cheap, renewable, clean energy. Yet by late 2005 only thirty-nine new nuclear power plants had been proposed by the nations of the world, and none were proposed for the United States. The percentage of electricity produced worldwide amounted to just 16 percent. The nuclear energy industry seemed to be stagnating (standing still; not moving forward).

Throughout the 1980s and 1990s and into the new millennium, the public began to have serious doubts about the safety of nuclear power. Those doubts arose because of the industry's first major setback, which took place on March 28, 1979. On that day an accident occurred at the Three Mile Island nuclear power plant

A nuclear accident occurred at Three Mile Island in 1979 which increased public awareness of some of the dangers of nuclear energy. © *W. Cody/Corbis.*

near Harrisburg, Pennsylvania. No one was injured or killed, and no one was overexposed to radiation from the plant. Still, the accident shut the plant down. If the accident had not been contained, a meltdown could have occurred. ("Meltdown" refers to an out-of-control reaction that overheats the reactor, causing it potentially to melt into the earth below, releasing radiation into groundwater and the atmosphere.) Many Americans started to distrust nuclear power, believing that the possibility of a catastrophe was too great. Not helping the industry was a major movie that year called *The China Syndrome*. The movie dramatized events at a fictional California nuclear power plant that were eerily similar to the Three Mile Island accident. Its title referred to the theoretical possibility that an overheated nuclear reactor could melt its way through the Earth to China.

Measuring Radiation

In measuring radiation and radiation exposure, physicists use a number of units of measurement, depending on exactly what they are trying to measure. Complicating matters is that there are "common units" of measurement and so-called "SI units," or "standard units." SI units are those recommended by the worldwide General Conference of Weights and Measures. Some of these units, such as curies (named after French physicists Pierre [1859–1906] and Marie [1867–1934] Curie) measure amounts of radiation. Others, such as rems, measure doses of radiation people might receive.

Then in 1986 a major disaster struck. On April 26 an explosion took place in reactor number 4 at the nuclear power plant in Chernobyl, a city in Ukraine (formerly part of the Soviet Union) about 70 miles (112 kilometers) north of Kiev. In this accident, a large amount of radiation was released into the atmosphere. Scientists estimate that the amount of this radiation was 100 to 150 million curies (although this unit is well known scientists now use the Bequerel as the unit of radiation), primarily in the form of radioactive cesium and iodine. Thirty-one people were killed in the accident, including firefighters, and 135,000 people within a 20-mile (32-kilometer) radius had to be permanently evacuated. Several years later, an additional 110,000 people were evacuated. Entire villages had to be decontaminated, and in the years that followed the rates of certain cancers among people in the area were noticeably higher. (Exposure to radiation increases the risk of developing cancer.) Radioactivity spread over large areas of the Soviet Union, into Eastern Europe, and as far away as Scandinavia. It is estimated that the accident cost the Soviet Union $12.8 billion. The human costs—stress, lost homes, poor health—cannot be measured.

These accidents burst the nuclear industry's bubble. People began to fear a major accident that would dwarf the kinds of accidents that took place at conventional coal-fired electric-generating stations. On December 16, 2005, the world held its breath when a large explosion damaged a Russian nuclear power plant outside the city of St. Petersburg.

The nuclear industry began to face other problems in the 1980s and beyond. The cost of building nuclear power plants was spiraling out of control. Most new plants went far over budget. Also in the 1980s and 1990s, the first aging nuclear plants had to be shut down and taken out of operation. It was discovered then that the cost of decommissioning (shutting down) a nuclear power plant was high because extreme care had to be taken to dispose of radioactive components properly. On top of these problems, the waste from nuclear power plants was beginning to accumulate, and no one knew quite what to do with it.

Because of these problems, plans for construction of new plants were in many cases canceled. By 2005 the number of operating plants in the United States had declined (to 103) as older plants were decommissioned. Nuclear power had become an emotional issue. Its supporters believe that by the year 2050, the energy needs of the United States will triple. They believe that other forms of alternative energy can help, but only nuclear plants can provide power on a large scale. Opponents of nuclear power, however, believe that the costs and the risks are too high.

HOW NUCLEAR ENERGY WORKS

Generating electricity through nuclear power is an enormously complex technical feat. It takes the combined skills of geologists (scientists who study Earth's structure, especially rocks), mine operators, engineers, and scientists, as well as large numbers of highly trained and skilled plant operators. The federal government oversees the construction and operation of these plants to make sure that they are built and operated to the very highest standards.

Uranium

Producing nuclear power begins with the fuel, uranium. Uranium was discovered in 1789 by a German chemist, Martin Klaproth (1743–1817). He discovered uranium in a mineral called pitchblende. The element was named after the planet Uranus, which had been discovered just eight years earlier. Scientists' best guess is that uranium was formed in supernovas (or exploding stars) about 6.6 billion years ago. In the Earth, radioactive decay of uranium is the planet's main source of internal heat.

Uranium is used primarily in the nuclear industry, but it has other uses as well. Because it is a dense, heavy element (18.7 times as dense as water), it is sometimes used in the keels of boats as a weight to keep them upright. (Density refers to weight relative to volume. A ton of feathers weighs as much as a ton of lead, but because lead is denser than feathers, it takes up far less volume.)

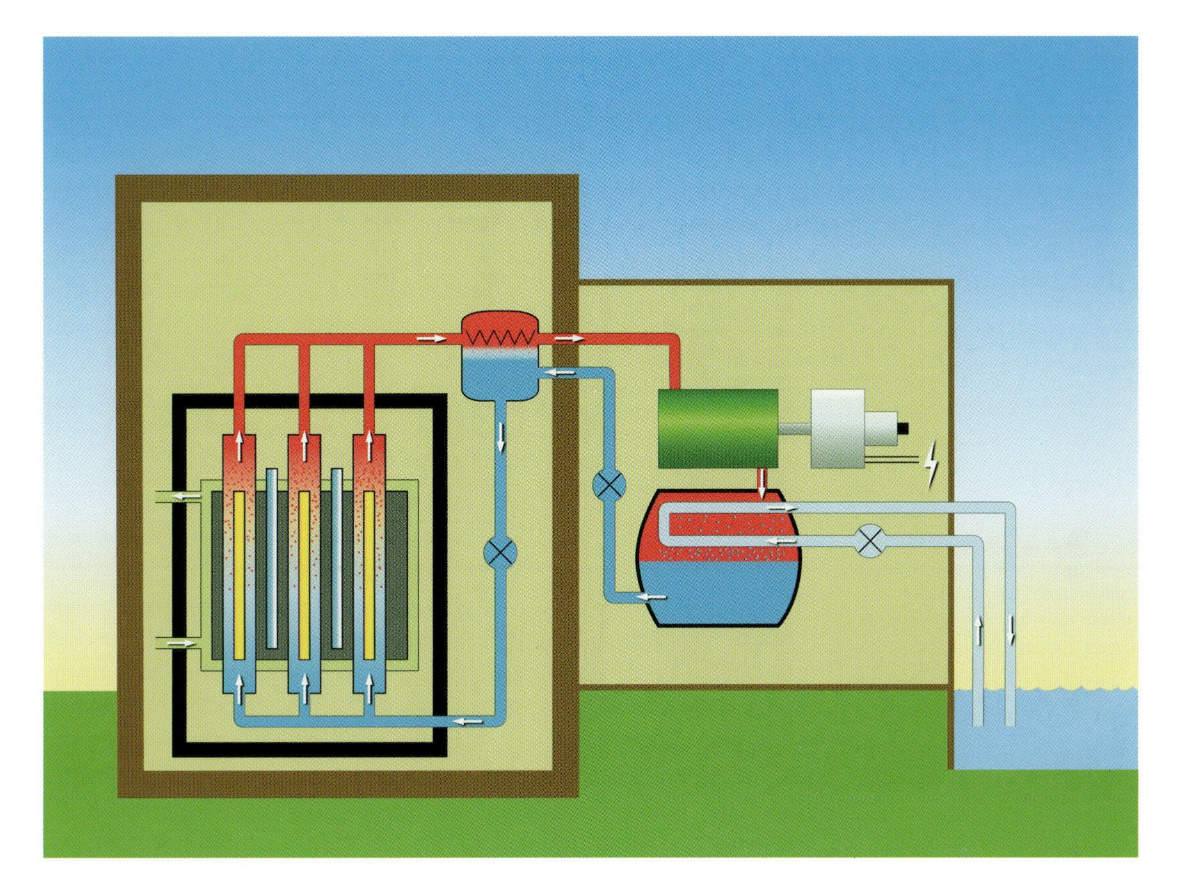

Diagram of the workings of an RBMK nuclear reactor, the type used in the Chernobyl power station. In this reactor, the core comprises fissile fuel rods (yellow) surrounded by water, encased in graphite. The water is heated by the reactions, producing steam (red). The steam passes through a moisture separator (upper center) and then to a turbine, which drives the electricity generator. The steam is condensed back to water by a cooling circuit. The flaw in this design is that power output increases with loss of cooling water. This was responsible for the 1986 Chernobyl disaster, which caused radioactive contamination of much of northern Europe. *SPL/Photo Researchers, Inc.*

Its density also makes it useful as a counterweight in such applications as airplane rudders, and it makes a good radiation shield.

The uranium atom

Uranium is the heaviest naturally occurring element. It has sixteen different isotopes, although the most common ones are U_{235} and U_{238}. U_{234} is found in trace amounts and results from the decay of U_{238}. The more abundant isotope, U_{238} (which accounts for 99.3 percent of the uranium in the Earth's crust) plays a role in keeping the Earth warm. Like any radioactive substance, U_{238} decays, but it decays very slowly. Its half-life is about the same as the age of the Earth, 4.5 billion years. ("Half-life" is a term scientists use to refer to the rate at which a radioactive substance decays, or breaks down. Thus, half of all U_{238} has broken down over the past 4.5 billion years. Half of the half that is left will break down over the next 4.5 billion years, and so on.) From the standpoint of nuclear energy, the important isotope of uranium is U_{235}.

The nucleus of a U_{235} atom consists of ninety-two protons and 143 neutrons. This is the isotope of uranium whose atoms can be split relatively easily. When a U_{235} atom is struck by a neutron, the atom splits, releasing energy. It also releases two or three neutrons of its own, which in turn split other atoms, and on and on in a chain reaction. In a nuclear reactor, the released energy is at first kinetic energy. Kinetic energy is the energy contained in anything (such as water, wind, or a neutron) that is in motion. But submicroscopic particles travel only tiny distances, so the kinetic energy is rapidly converted to heat (similar to the way the brakes on a car get hot when they stop the kinetic energy of a moving car). This heat is then used to produce steam, which turns a generator to produce electricity. Heat makes up about 85 percent of the energy released. Most of the rest of the energy is in the form of gamma rays. (A gamma ray is a photon that is released by a radioactive substance. A photon is a form of energy, like light.)

In many respects, the process as described is much more complex. For example, physicists note that only isotopes with an odd number of particles in the nucleus, like U_{235}, are fissile (able to be split). Further, not every neutron that hits a uranium atom causes fission. Sometimes the neutrons are absorbed by the atoms they strike, so no fission takes place. Other neutrons simply escape and do nothing. Another complication has to do with the speed of the neutrons. Some are called "prompt neutrons," but others experience a delay of up to 56 seconds.

The challenge for nuclear engineers is to keep the ongoing fission reaction in precise balance. When the reaction is in balance, scientists say that it has reached "criticality." At criticality, the neutrons are doing their work in balance, meaning that their numbers remain constant and under control. The pace of the reaction can be speeded up or slowed down by increasing or decreasing the number of neutrons. If the increase is too rapid, the reaction can almost instantaneously get out of control.

Plutonium

Plutonium (chemical symbol Pu), named after the planet Pluto, is an element that forms in a reactor core as the isotope Pu_{239}. It forms when U_{238}, which is also present in nuclear fuel, absorbs a neutron. Now the atom has an odd number of particles in the nucleus, making it fissile in the same way that U_{235} is. But like U_{235}, it sometimes just absorbs the neutron, creating the isotope Pu_{240}, which is not fissile. Over time, the amount of Pu_{240} builds up in the fuel rods. When the rods are "spent," or no longer usable as fuel, this plutonium can be recycled. It undergoes a conversion process that makes it usable as nuclear fuel. Not all nuclear reactors are designed to allow this recovery and conversion process. Those that do are called "breeder reactors," for they "breed," or produce, additional fuel.

Plutonium is perhaps the most highly toxic substance that exists. The smallest amount can cause such diseases as lung cancer. Workers who handle plutonium observe the strictest safeguards to avoid exposure.

Uranium: From the ground to the reactor

While uranium can be found in seawater, it is found most commonly in rocks and is as common as the elements tin and gold. It exists in concentrations of about two to four parts per million. Uranium is mined in at least two ways. One is to dig up the ore that contains it, crush the ore, and then treat it with acid, which dissolves the uranium to remove it from the ore. The other is a process called in situ leaching (*in situ* is Latin for "in place"). In this process, the uranium is dissolved from rock and pumped to the surface of the Earth. Either way, the end result is a compound called uranium oxide, or U_3O_8. This material is often referred to as "yellowcake."

The uranium, though, cannot be used as fuel in this form. It first has to be "enriched," so mine operators sell the yellowcake to uranium enrichment plants. The first step in converting it into a usable fuel is to convert it into a gas, uranium hexafluoride, or UF_6. This increases the amount of uranium from its natural level of 0.7 percent to 3 to 4 percent, so the uranium is said to be "enriched." The next step is to convert the uranium hexafluoride to uranium dioxide, or UO_2. Uranium dioxide can then be processed into pellets that are about the size of a knuckle on a person's finger. The pellets are then inserted into thin, 12-foot-long (3.5-meter-long) metal tubes, called fuel rods. Bundles of these tubes are then inserted underwater into the core of the nuclear reactor.

Inside the reactor

A nuclear power plant has been constructed, probably at a cost of anywhere from $3 billion to $5 billion, or even more. Construction of the plant took at least four years, possibly up to ten years. Geologists have carefully considered the site of the plant to make sure that the chances of it being damaged by an earthquake or volcanic activity are small. Engineers and construction workers have carefully built the plant. The materials used were of the highest standards. Every weld in metal components was closely examined and even x-rayed to be sure it is as close to perfect as possible. Provisions were made to ensure that the plant is secure, so that terrorists or others cannot enter and take it over. Provisions have also been made for the safety of the plant's employees so they can quickly shield themselves from radiation in the event of an accident. The plant is built with "redundant," or repetitive, safety systems, so that if something breaks down, there is a backup. The most critical of these systems is water that can be used to cool an overheated reactor. No detail is overlooked.

As the time approaches for the plant to come online and begin producing power, the fuel is inserted into the tubes and the tubes, up to 200 of them, are inserted into the reactor core. Then, at the appropriate moment, the control rods are slowly pulled out. These rods are generally made of graphite or boron, and they control the pace of the nuclear reaction by absorbing neutrons. The farther the rods are inserted, the more neutrons they absorb, slowing down or stopping the reaction. As they are withdrawn, more and more neutrons make it to their target, and the chain reaction begins.

At this point the plant is nowhere near ready to operate at maximum power output. For weeks, the plant's engineers will fire up the reactor very slowly. They will check and recheck every component of the plant to make sure that everything is operating properly and safely. After a period of several weeks of testing, the reactor will begin producing power at its normal operating level, and consumers will begin enjoying the benefits of the electricity it produces.

CURRENT AND FUTURE TECHNOLOGY

Nuclear power plants come in many different shapes and designs. Many of the first plants to be constructed were huge, enabling them to produce the greatest amount of power possible. More recent designs are smaller, making them less costly and easier to build. But despite their many technical and engineering differences, nuclear reactors come in two basic types: pressurized water systems and boiling water systems.

Pressurized water reactor system

One system in common use is called the pressurized water reactor system. It is given this name because it relies on water under pressure to produce the heat needed to produce electricity. In such a system, the fuel rods are inserted into a steel pressure tank that contains ordinary water. The water acts as a coolant, but it also moderates the reaction because it can absorb neutrons. Protruding (sticking out) through the lid of the pressure tank are the control rods.

As the control rods are slowly pulled out, the chain reaction begins. The reaction produces heat, which heats the water in the pressure tank. The water heats to 518° Fahrenheit (270° Celsius). The water does not boil, though, because it is under intense pressure.

The heated water is then channeled to a heat exchanger in a closed circuit. The water in the heat exchanger is then heated up, producing steam. The steam drives a turbine generator that is little different in principle from a turbine used in a windmill or a hydroelectric dam. As the generator turns, it produces electricity. Meanwhile, the steam is condensed, usually by cool water from a lake or river, and returned to the heat exchanger.

Boiling water reactor system

The other major system, the boiling water reactor system, is more efficient than the pressurized water system. One noticeable difference is that with a boiling water system, the control rods protrude from the bottom of the containment chamber. Inside the chamber is the reactor core. The control rods are at the bottom because the water inside the chamber is allowed to boil. The steam created by the boiling water is allowed to rise to the top of the chamber. Pipelines carry the steam directly to the turbines, where its heat causes them to turn to create electricity. The steam then condenses and is channeled back into the containment chamber. Underneath the reactor is a circular tunnel filled partway with water. This tunnel is a safety mechanism. If any steam or water were to escape from the containment chamber, it would fall into the tunnel, where it could do no immediate harm.

The possibility of nuclear fusion

Scientists look forward to the discovery of a power source that is clean, safe, universally available at all times to all people throughout the world, and that uses a fuel that is abundant, cheap, and efficient. It would not contribute to global warming or air pollution, require large plants that would disrupt the natural environment, or produce dangerous by-products. To that end, some scientists conduct research into what is called "cold fusion." Cold fusion uses fuel that is commonly available from the hydrogen in water. However, governments have favored a more conventional approach to fusion at extremely high temperatures. In 2005, Cadarache in

France was chosen as the site for the International Thermonuclear Experimental Reactor. This will be built as a cooperative venture between the EU, U.S., Russia, China, Japan, and South Korea. This is a major step in the development of fusion as a potential large-scale source of electricity that will not contribute to climate change.

Nuclear *fission* refers to the splitting, or breaking apart, of atoms. Nuclear *fusion*, as the name suggests, involves the fusing, or joining together, of atoms. The light nuclei of two atoms bind together during nuclear fusion to form a single heavier nucleus. One example is the deuteron, a single particle formed by the combination of a neutron and a proton. When a deuteron or similar particle is formed, its mass is generally less than the total mass of the two original particles. The mass that disappears is released as energy. What appeals to scientists seeking to harness nuclear fusion is that such reactions occur in nature throughout the universe, particularly in stars. Fusion takes place in stars because of their high temperatures, up to 18,000,032° Fahrenheit (10 million° Celsius), possibly even hundreds of millions of degrees. The problem is that while such high temperatures can be found in the center of stars, including the Earth's sun, they do not occur naturally on Earth.

Despite the high temperature needed for fusion to occur, scientists have tried to reproduce fusion reactions on Earth. The process they formulated was to use two isotopes of hydrogen. These isotopes, called "heavy hydrogen" because they contain extra atomic particles, are deuterium and tritium. While a normal hydrogen atom consists of a single electron and a single proton in the nucleus, deuterium also contains one neutron in the nucleus and tritium contains two. These isotopes fuse at lower temperatures than do the nuclei of regular hydrogen atoms, and they are relatively abundant. In the oceans, about one in 6,500 or 7,000 hydrogen atoms are deuterium, and they can be easily extracted. The source of tritium is an element called lithium, which is abundant in the Earth's crust.

Scientists discovered that when a mixture of deuterium and tritium is raised to a high enough temperature, or when the elements are accelerated to a very high speed, one deuterium nucleus fuses with one tritium nucleus. The result is a new element, helium. More importantly, excess energy is given off in the form of a neutron that moves at a very high speed. Scientists believe that fusion could be the "fuel of the future" because the fuel—deuterium and tritium—contains an enormous amount of energy, called "density" by scientists. It has been estimated that a single thimbleful of heavy hydrogen contains the same amount of energy as 20 tons of coal. An amount that would fill the bed of a pickup truck would provide the same amount of energy as 21,000 rail cars full of coal or 10 million barrels of oil. Further, using such fuel would be extremely safe. The only by-product is helium, and there is no danger of a fusion reaction spinning out of control. If the fuel escapes, the fusion reaction simply stops.

So far, fusion experiments have failed to produce any power in excess of the power needed to produce the fusion reaction. In other words, there was a net power loss. For many scientists, the enormous energy demands of hot fusion make it impractical. Instead, they have searched for a way to create fusion reactions at low temperatures, called "cold fusion." Cold fusion is a term coined in 1986 by Dr. Paul Palmer of Brigham Young University in Utah. It is the popular term for what scientists call "low energy nuclear reactions" in a field that is sometimes called "condensed matter nuclear science."

In 1984 two scientists, Stanley Pons of the University of Utah and Martin Fleischmann from England's University of Southampton, began conducting cold fusion experiments at the University of Utah. On March 23, 1989, Pons and Fleischmann made an announcement that startled the world. The two claimed that they had successfully carried out a cold fusion experiment. This experiment produced excess heat that could be explained only by a fusion reaction, not by chemical processes. Many scientists, though, disputed their claim. They tried to duplicate the Pons-Fleischmann experiment and failed.

So the question remains: Is cold fusion possible? Some scientists answer with a no. Many other scientists, though, disagree. They point out that cold fusion research is still just beginning. Some of the problems reported with duplicating the Pons-Fleischmann findings have been the result of normal uncertainties about how to design and conduct experiments to get consistent results.

Meanwhile, many scientists have made claims that they have produced cold fusion. Some of the most prominent researchers in the field are in Japan, where the level of funding for cold fusion research is much higher than it is in the United States. At Japan's Hokkaido University, for example, D. T. Munzo reported experiments in which the ratio of energy output to energy input was seventy thousand to one. As of 2005, though, the world seemed decades away from seeing a commercial fusion reactor, whether hot or cold.

In 1991 Greenpeace activists placed some 3,000 wooden crosses next to the Chernobyl nuclear power plant, commemorating the nuclear disaster five years earlier. ©.*Reuters/Corbis.*

BENEFITS AND DRAWBACKS

In the imaginations of many people, nuclear power plants are surrounded by a field of radiation. As they drive down the highway and see the characteristic cooling tower of a nuclear power plant rising on the horizon, some people feel a slight twinge of anxiety. They know that they are not being exposed to radiation, yet their emotions make them wonder whether maybe they are.

Supporters of nuclear energy dismiss these concerns. They argue that nuclear power plants are safe and that nuclear power offers many significant benefits. At the same time, nuclear power has significant drawbacks, particularly the potential for accidents, the problem of nuclear waste disposal, and the possibility that terrorists could attack nuclear power plants.

Benefits

The benefits of nuclear energy include the following:

1. Many scientists believe that nuclear energy remains the best way to provide large amounts of power for a large and growing world population. A typical nuclear power plant produces 1,000 megawatts, or 1 billion watts, of electricity. Other forms of alternative energy produce far less, particularly relative to their size. For example, the largest wind farm in the United States is the Stateline Wind Energy Center along the Columbia River on the Washington-Oregon border. This massive farm consists of 454 wind turbines, each 166 feet (50 meters) tall and, at peak capacity, generating 660 kilowatts, or 660,000 watts of power. Because of changing wind conditions, the windmills do not always operate at peak capacity. To provide power equivalent to that of nuclear power plants, immense numbers of large wind farms would have to be built.
2. Nuclear energy is reliable. In contrast to most other forms of alternative energy, nuclear energy can be provided on a consistent, predictable basis nearly anywhere in the world. It is not subject to weather conditions. In contrast, solar power requires consistent sunshine, so not all areas are suitable for solar power. Wind power has similar limitations. Hydroelectric dams provide large amounts of power worldwide, but the number of rivers that remain suitable for damming is limited. Such alternatives as ocean wave power and tidal power are likewise limited by geography and unpredictable weather patterns.
3. The supply of fuel for nuclear power is abundant. Uranium exists throughout the Earth's crust, although in some places, it can be mined more easily than in others. Scientists estimate that the amount of uranium known to be readily available is enough to last fifty years. However, they also point out that its relative abundance has not made it necessary for mining companies to search very hard for it. Scientists are confident that more intensive searching will yield abundant new reserves of uranium. While uranium is not renewable, as wind and solar power are, enough probably exists for many centuries to come. Further, nuclear plants produce plutonium as a by-product of the nuclear reaction. This plutonium can be reprocessed into fuel.

4. The price of nuclear fuel remains relatively constant, and its sources remain relatively consistent. Uranium is mined extensively in about twenty countries throughout the world. The relatively large number of suppliers ensures that prices do not change rapidly and unexpectedly. In contrast, the world's petroleum reserves are in the hands of a small number of countries. Many of these countries are politically unstable. As the Arab oil embargoes of the 1970s showed, oil supplies to the United States and other countries can be cut off overnight for political reasons. Uranium is not subject to these uncertainties, and nations such as the United States and Canada can mine their own uranium. In fact, Canada leads the world in uranium mining. Another leading producer is Australia, which, ironically, has no nuclear power plants.
5. Nuclear power plants have a low impact on the environment. A chief advantage of nuclear power is that it does not require the burning of fossil fuels such as coal. Thus, it is cleaner than fossil fuels and does not contribute to pollution.
6. Nuclear power plants are safe. As of late 2005 the only deaths that have ever resulted from a nuclear power plant accident occurred at the Chernobyl plant in Ukraine. Nuclear experts, though, note that the design of the Chernobyl plant was extremely outdated and that the plant was not very well constructed. This was a common problem for all types of construction under the Communist regime of the old Soviet Union. They believe that the kind of accident that happened at Chernobyl is much less likely with more modern and better built plants. This has meant that despite worries among the public, politicians have increasingly seen modern nuclear reactors as a source of energy that avoids emission of greenhouse gases and after a period where few reactors have been built they are being re-considered as energy sources.
7. With regard to safety, the track record of the nuclear industry has improved over the 1990s and early 2000s. For example, when something in the operation of a nuclear plant gets out of kilter, a "scram" takes places. This refers to a wide range of automatic safety mechanisms. Alarms sound, backup systems kick in if necessary, and the plant's controls automatically make necessary adjustments, particularly making sure that water surrounds the reactor

core to keep its temperature under control. If necessary, the nuclear reaction stops and the reactor shuts down. The nuclear industry keeps track of the number of scrams per 7,000 hours of operation, or about one year. In the late 1990s two-thirds of U.S. nuclear power plants had zero scrams. The number of scrams at the other third was extremely low, and usually the problems that caused them were minor and easily fixed.

8. A major concern for nuclear plant workers is exposure to radiation. People are exposed to radiation every day of their lives. Radiation reaches the Earth from the sun, and it radiates from rocks in the earth. This radiation is referred to as "background radiation," and it varies with altitude (height above sea level) and geography. People in such countries as Finland are exposed to three times as much background radiation as Australians. Even on an airline flight over the North Pole from, say, Tokyo to London, people are exposed to cosmic radiation seven to eight times the normal level.

Drawbacks

Despite its many benefits, nuclear power has significant drawbacks as well. Throughout the 1990s and into the new millennium, scientists, environmentalists, and the public have focused more of their attention on these drawbacks. As a result, nuclear power has become an emotional political issue. Its opponents are passionate in their belief that nuclear power poses a significant danger to the world. Some of their concerns include the following.

Catastrophic accident

The potential for a catastrophic accident continues to exist. The world's nuclear power plants have accumulated a total of about twelve thousand years of operation. During that time, there have been only two significant accidents, Three Mile Island (although the public was not exposed to radiation during that accident) and Chernobyl. Supporters of nuclear power point out that far more people lose their lives in accidents at conventional power plants in one year than have lost their lives in nuclear accidents.

The problem is one of public attitudes rather than statistics. Opponents of nuclear power note that a catastrophic accident at a conventional power plant might be tragic for those injured and killed. Still, the effects would be limited to the plant itself and perhaps the immediately surrounding area. Deadly radiation would not be released into the atmosphere. People would not have

Safety Stats

In 1998 the U.S. Bureau of Labor Statistics reported that for every 200,000 hours of work performed in nuclear plants, there were 0.34 accidents that resulted in injury. In contrast, for all other industries, the number was seven times greater, or 2.3 accidents per 200,000 worker hours.

to be evacuated, and those nearby when the accident occurred would not suffer the ill effects of radiation.

In contrast, a catastrophic accident at a nuclear plant could have enormous effects on the surrounding environment, effects that would last for decades, if not longer. Nuclear opponents believe that the risk is simply too great. One mistake, one faulty component, one operator error could create an environmental catastrophe. The margin for error is nearly zero. While the risk of a nuclear catastrophe is low, such a catastrophe would have high consequences.

Adding to the problem is the mysteriousness of anything nuclear. Ever since the atomic bombings of Japan at the end of the Second World War, people have been afraid of nuclear power. Excessive exposure to nuclear radiation can cause cancer, another word people respond to with fear. Few people understand nuclear physics. That sense of awe and mystery spills over into fear of anything "nuclear," including nuclear power plants.

Waste storage and disposal

Nuclear waste comes in two types: low-level and high-level. Low-level waste is produced by hospitals, which use radioactive materials for certain medical tests. Similar low-level waste is also used for research purposes at universities and other research facilities. This material has to be disposed of safely, and if it is done so, it poses little health risk to the public. The radioactivity in these materials breaks down quickly (usually in days or at most weeks), and the material can then be disposed of as normal trash.

High-level nuclear waste, such as that produced by nuclear power plants and in producing and dismantling (taking apart) nuclear weapons, is another matter. As of 2003 the United States had accumulated about 49,000 metric tons (a metric ton is about 2,200 pounds) of spent nuclear fuel rods. These are fuel rods that have been removed from power plants because the fuel is depleted.

A steel and concrete tube holding over 600 tons of nuclear waste sits in a secured holding area along the Pacific Ocean at the San Onofre Nuclear Power Plant near San Clemente, Calif. Storage of nuclear waste contintues to be a controversal issue. *AP Images.*

This amount would cover a football field to a height of 10 feet (3 meters). The U.S. Department of Energy estimates that the amount will total 105,000 metric tons by the year 2035. Much of this material is stored in water pools on the sites of nuclear power plants. No one knows what to do with this accumulating waste.

The problem with nuclear waste is the half-life of such elements as uranium and plutonium, as well as other radioactive materials produced in nuclear power reactors as by-products. Some of these by-products include cesium-137 and strontium-90, both highly radioactive. Most of these elements have extremely long half-lives. The half-life of plutonium is 24,000 years. The half-lives of some other radioactive elements are 100,000 years, even longer. This means that nuclear waste disposal has to be thought of in terms of geologic time, not next year or even next century. The ancient Roman Empire was thriving just 2,000 years ago; the ancient Egyptians, 3,000 years ago. Humans find it hard to think that far ahead.

Roughly every twelve to eighteen months, a nuclear plant has to shut down and all the fuel rods have to be replaced. These fuel rods are highly radioactive, so they cannot simply be taken to the nearest landfill. Strict precautions have to be taken to make sure that the spent rods do not pose a risk to the environment or to the public. Further, when a nuclear plant is "decommissioned," or shut down, the radioactive components in the core have to be disposed of properly. All of this is a difficult technical undertaking and one that carries a high expense.

Several proposals have been made for ways to dispose of high-level nuclear waste. One proposal is to launch it into space. Others are to bury it on a remote island or in the polar ice sheets. So far, these have not been attempted. Another proposal is to bury the waste under the seabeds. While technically possible, the expense of doing so would be enormous.

The most widely accepted possibility is to bury nuclear waste underground in stable geological formations. The waste would undergo first a process called vitrification (from the Latin word *vitrium,* meaning "glass"). This means that the waste is mixed with silica (like sand) and melted into glass beads. This process makes the waste more stable and reduces the chance that radiation could seep out into the air or water. The beads are then buried in an area that is geologically stable (that is, it does not experience earthquakes, tremors, or volcanic activity). When the storage facility is full, it would be sealed with rock.

The problem with this method is that no community wants to be home to the storage site. Nuclear waste would have to be trucked in, with the potential for accidents. Then the nuclear waste would be stored nearby, essentially forever. In 1983 President Ronald Reagan signed into law the Nuclear Waste Disposal Act. Under the act, the federal government took on responsibility for nuclear waste disposal. The act required the U.S. Department of Energy to find a suitable site for underground storage, then build the facility. In 2002 the department identified Yucca Mountain in Nevada as the most suitable site. Understandably, Nevadans do not want to be the dumping ground for the nation's nuclear industry and have opposed this plan. The state's governor notified the federal government that Nevada opposed the plan. The U.S. Congress voted to override the governor's objections. Accordingly, the federal government has designated the Yucca Mountain site as a long-term storage facility for about 70,000 metric tons of nuclear waste. As of late 2005, however, the issue was still not entirely resolved. No steps had been taken to construct the facility.

Another problem the nuclear industry has created is "mill tailings." These are waste materials created in mining uranium ore.

A worker walks down the tunnel almost half a mile inside Yucca Mountain, where the U.S. Department of Energy hopes to store the nation's high level nuclear waste. © *Dan Lamont/Corbis.*

The materials contain trace amounts of uranium left behind, as well as radium and thorium, both radioactive. The radioactive material cannot simply be left in place. The federal government, specifically the U.S. Nuclear Regulatory Commission, regulates the removal, storage, and monitoring of mill tailings.

Terrorism

After the terrorist attacks on the United States on September 11, 2001, policy makers raised concerns about the security of the nation's nuclear power plants. It is known that members of al-Qaeda, the Islamic terrorist network, have been instructed and trained in ways to attack power plants. The concerns of policy makers and nuclear regulatory officials are many:

As they did on September 11, terrorists could hijack an airliner and fly it into a nuclear power plant. The scientific director of the Nuclear Control Institute believes that a direct, high-speed impact by a large airliner "would in fact have a high likelihood of penetrating a containment building" with a nuclear reactor inside. "Following such an assault," he said,

Why Yucca Mountain?

The federal government identified Yucca Mountain, about 100 miles (161 kilometers) northwest of Las Vegas, as the best site in the United States for long-term nuclear waste disposal. This site was selected for a number of reasons that highlight the problems of disposing of nuclear waste:

The area has a dry climate. Yucca Mountain receives only about 7.5 inches (19 centimeters) of rainfall each year. Most of the rain runs off or evaporates. The rainfall that remains moves through the rock at a rate of only about .5 inch (1.27 centimeters) per year.

Yucca Mountain is stable geologically. Studies have shown that Yucca Mountain has not changed much for at least one million years. The earth surrounding the mountain does not shift because of volcanoes or earthquakes. Because the waste would be 1,000 feet (305 meters) below the surface, any earthquakes that did take place would likely not allow any of the material to leak out. This is because earthquakes are most intense at the Earth's surface.

The Yucca Mountain site has a deep water table. The water table, the level at which underground water is reached, is about 2,000 feet (610 meters) below the surface. The nuclear waste would be stored about 1,000 feet (305 meters) below the surface. Therefore, the water would

"the possibility of an unmitigated [unstopped] loss-of-coolant accident and significant release of radiation into the environment is a very real one." Other scientists believe that most nuclear plants could withstand the impact of an airliner.

Terrorists could steal plutonium or highly enriched uranium, either from the plants themselves of from uranium enrichment facilities. It takes only about 18 pounds (8 kilograms) of plutonium or 55 pounds (25 kilograms) of highly enriched uranium to build a nuclear weapon. But in the nuclear industry, these materials are moved about by the ton, and accurate records are not always kept. Policy makers believe that a sophisticated terrorist group could steal these materials and make a nuclear bomb. The materials could also be used to construct so-called "dirty bombs," or what experts call never reach the waste. If by some chance it ever did, the water that flows under Yucca Mountain continues to flow underground into Death Valley, a forbidding desert. None of this water is used to supply water to nearby cities. Further, the Yucca Mountain site is in an enclosed water basin. This means that the area is completely surrounded by higher land. This in turn means that water flows downward and stays put. It does not spill into aquifers (water-bearing rock and sand) that supply drinking water.

The area is in a remote location. No one lives on Yucca Mountain, and the nearest people are 15 miles (24 kilometers) away. Most of the land around Yucca Mountain, about 1,375 square miles (3,561 square kilometers), has been taken over by the federal government. It is also on the edge of sites that were once used to test nuclear weapons, sites on which no one wants to live or work. If that area is added in, the unpopulated area is 5,470 square miles (14,167 kilometers).

Finally, access to the Yucca Mountain site is highly restricted. The U.S. Air Force maintains training sites and gunnery ranges in the area. The area is dense with security personnel and procedures, so it would be nearly impossible for anyone to disturb the site. Further, geologists have determined that the site has no valuable minerals, oil, precious metals, or other assets. Therefore, geologists believe that, even thousands of years from now, no one would have any reason to dig the site up.

"radiation dispersal devices." These are bombs made of conventional explosives such as dynamite that are packed with nuclear materials, even nuclear waste. The explosion would disperse, or distribute, the radioactive materials around a wide area. The result would be public panic and an area contaminated with radiation.

Policy makers are also concerned about security at nuclear facilities. After September 11, training exercises were carried out at nuclear plants to see how well the plants' personnel could resist a terrorist attack. Military personnel disguised as terrorists attempted to gain access to these plants. Some experts claim that at nearly one-half of U.S. nuclear power plants, armed guards were not able to stop these mock attacks.

A final concern is nuclear proliferation. *Proliferation* means "spreading," and the concern is that nations can develop nuclear power—or claim to—and convert their nuclear capabilities into weapons. In 2005 many nations of the world, including the United States, were opposing nuclear development programs in Communist North Korea and Iran. While these countries insisted that their programs were for peaceful purposes, worries persisted that they were trying to develop nuclear weapons.

ENVIRONMENTAL IMPACT

The chief benefit to the environment of nuclear power plants is that they do not emit (give off) harmful gases, such as carbon dioxide and sulfur dioxide. In this way they differ from conventional power plants, which emit these gases primarily because they burn coal, a fossil fuel. If the energy generated by nuclear power plants worldwide were instead generated by burning coal, the amount of additional carbon dioxide released into the atmosphere would be about 1,600 million tons. Moreover, burning coal releases toxic heavy metals, including arsenic, cadmium, lead, and mercury. Nuclear energy prevents release into the atmosphere of about 90,000 tons of these metals each year. France's heavy reliance on nuclear power has lowered that country's air pollution from electrical generation by 80 to 90 percent.

By not emitting these gases, nuclear energy does not contribute to environmental problems such as air pollution, smog, and the "greenhouse effect." The greenhouse effect refers to the ability of some gases, such as carbon dioxide, to accumulate in the air. The theory is that in doing so, they act like a greenhouse, trapping the sun's heat. In turn, many scientists believe that this trapped heat is increasing average temperatures around the world. This increase is referred to as "global warming." Global warming is blamed for the melting of the polar ice, raising sea levels and endangering coastal cities. (Not all scientists agree that this is happening.) Further, by not emitting pollutants, nuclear power plants do not contribute to acid rain. Acid rain is any form of precipitation that is more acidic than normal because the water has absorbed acidic pollutants from the air. Acid rain can harm crops and forests. It can also contribute to the deterioration of buildings and public monuments, which dissolve because of the acid in precipitation.

Nuclear power plants also do not harm surrounding bodies of water. A myth that some people believe is that nuclear plants discharge water into nearby lakes and streams that is either radioactive or extremely hot. This is not true. The water released from a nuclear plant never comes into contact with the radiation. Further, if the water is too hot to be discharged, it is cooled either in a cooling pond or in cooling towers before release.

Supporters of nuclear power point out that some other alternative forms of energy do not have the same low impact on the environment, especially hydroelectric dams. While such dams have the benefit of not emitting harmful gases or pollutants, the dams have a major impact on the surrounding environment. By turning rivers into huge lakes, they disrupt vegetation and wildlife. Many dams have displaced (driven out) large numbers of people. Further, the reservoirs behind hydroelectric dams emit their own form of pollution. As the water level of the reservoir falls, the wet ground that surrounds it supports the growth of vegetation. As the water rises, this vegetation is covered and rots. The rotting vegetation emits methane gas, a pollutant. In addition, hydroelectric dams have an adverse effect on fish because they disrupt breeding and spawning grounds.

Nuclear power plants do not have harmful effects on wildlife. In fact, they often can have beneficial effects. For example, when cooled water is released from the plant, the water often contributes to the formation of wetlands. These wetlands can become nesting grounds and provide habitat for birds, fish, and other animals. Some companies that build and run nuclear plants even develop wildlife preserves and parks in the surrounding area, where plants grow abundantly in the moist soil.

Even species that are endangered (that is, in danger of becoming extinct) have found new life around nuclear plants. Some of these species thrive nearby, including such endangered species as bald eagles, red-cockaded woodpeckers, peregrine falcons, osprey, and the beach tiger beetle. The areas around nuclear plants are also home to such nonendangered species as wild turkeys, sea lions, bluebirds, kestrels, wood ducks, and pheasant.

Again, supporters of nuclear power point out that other forms of alternative energy do not have the same benefits. They agree that solar power and wind power are cleaner forms of energy, but they require huge "farms" of solar panels or windmills to produce significant amounts of electricity. Some argue that wind farms hurt an area's bird populations because the birds become almost hypnotized by the turning blades and fly right into them, where they are killed. By reducing an area's bird populations, the rodents that birds eat can multiply freely and cause rodent infestations.

Nuclear power protects land and animal habitats. Per unit of electricity, nuclear power plants take up far less land than other types of power-generating stations. For example, assume a plant that produces 1,000 megawatts of power (a megawatt is a million watts, so a thousand megawatts is 1 billion watts). To produce the same amount of power, a solar "farm" would need 35,000 acres of solar panels. A wind farm would require 135,000 acres devoted to windmills. In contrast, a typical nuclear power plant takes up only about 500 acres of land.

Further, the fuel nuclear power plants use, uranium, is very energy dense. This means that a pound of the fuel produces far more energy than a pound of coal. For example, one metric ton of uranium, or about 2,200 pounds (998 kilograms), will power a 1,000-megawatt nuclear power plant for two weeks. This fuel would come from about nine metric tons of mined uranium oxide. The same amount of energy from coal would require about 160,000 metric tons, or almost 353 million pounds. Thus, mining nuclear fuel has much less impact on the environment.

With regard to energy output, some nuclear power opponents say that these figures are misleading. They point out that conventional fuels like coal have to be burned to process uranium for use as fuel. They are correct, but the amount of conventional energy that has to be burned to do so is about 2 percent of the amount of energy the uranium will produce.

ECONOMIC IMPACT

In examining the cost of nuclear energy, many factors have to be taken into account. Some of these are obvious, such as construction costs and the cost of mining uranium. Others are more hidden and include taxes, licensing fees, interest payments on debt, and the like. Thus, any examination of the economic costs and benefits of nuclear energy involves complex calculations.

The first cost comparison involves the fuel itself. Uranium has to be mined, converted, enriched, and loaded into fuel rods. Coal has to be mined, but it can be used as is. On the other hand, the cost of transporting nuclear fuel is low because of its energy density. The cost of transporting coal is high because large volumes have to be shipped.

Per unit of energy, the cost of a nuclear power plant is generally higher than that of a conventional power plant. Nuclear power plants have to be built to the highest standards. Many of their systems are redundant, or repetitive, for safety reasons. On the

The Pacific Pintail, transporting 140 kg of weapons-grade plutonium, docks at Cherbourg, France after arriving from the United States on October 6, 2004. The nuclear waste will be conditioned here before being transported from this northwestern French port some 745 miles (1,200 kilometers) by road to a plutonium fuel fabrication facility in Cadarache, southern France. ©*Jacky Naegelen/Reuters/Corbis.*

other hand, coal-fired plants have additional costs because of requirements that they have pollution-control devices, such as scrubbers that remove particles from their emissions. Debt also increases the cost of nuclear plants. Because building these plants is so expensive, power companies have to borrow large sums of money, and they have to pay interest on that debt. Thus, high interest payments have added to their costs.

Nuclear power plants have higher maintenance costs than do conventional power plants. For example, corrosion and cracking are common problems in the water pipes in boiling water reactors. These components have to be replaced at great cost. In the meantime, the reactor is shut down. It is not producing energy, but workers still have to be paid and debt still has to be financed.

Both conventional and nuclear power plants have normal day-to-day costs. Nuclear facilities require highly trained technicians, engineers, safety inspectors, health workers, and the like, increasing labor costs. Conventional plants are relatively simple to operate, so they do not require as many highly trained workers. However, they require a larger labor force because of the amount of labor involved in running the plant's operations.

Nuclear plants face other charges as well. The license fee for a nuclear reactor is almost $3 million. The license for nuclear fuel use is over $2.5 million. Many nuclear plants pay $15 to $20 million in local property taxes. In addition, the Nuclear Regulatory Commission requires nuclear plant operators to take on expenses for other specialized needs, including, for example, radiographers who measure radiation in the plant. Producing nuclear energy does not come cheap.

On top of all these expenses, the nuclear industry spends many dollars for nuclear waste disposal. Coal-fired plants have only to dispose of ash. Further, the cost of decommissioning a nuclear power plant is high, often 4 percent of the initial cost of construction. A coal-fired plant that is put out of commission essentially just has to be knocked down and carted away. Yet most costs are comparable. The end result is that nuclear power is slightly more expensive than coal.

SOCIETAL IMPACT

The societal impact of nuclear power tends to be a matter more of perceptions and public sentiment than facts. Opinions about nuclear power are likely to depend on opinions about science. On the one hand, many people place a great deal of faith in science. They believe that science can solve many of the world's ills. Science, for example, can increase crop yields in poorer nations. It can reduce and eventually eliminate many diseases. And it can provide for the energy needs of the six billion people who live on Earth—a number that is likely to grow significantly as the twenty-first century progresses. Scientists, with their specialized knowledge, have become almost like magicians who solve the world's problems.

As the sheer volume of scientific information grows each year, however, the public feels disconnected from scientists and their magic. Few people know how a toaster works, let alone something as complex as a nuclear power plant. Further, they believe that while science can solve problems, it has also caused problems. In their view, the Earth and its resources have been exploited in the name of science. The atmosphere and bodies of water have been polluted because of scientific and technological advancement. Some of the people who feel this way yearn for a simpler time, when people (in their view) lived in harmony with the natural world. They were attuned to the cycles of the natural world and accepted them rather than trying to conquer them through science.

Nuclear energy stands at the center of this dilemma. Supporters of nuclear energy point to its clear benefits. It provides large amounts of power. It does not release pollution into the atmosphere. It does not consume resources whose supply will eventually run out. It does not make countries such as the United States dependent on foreign sources of fuel. It has an exemplary safety record, and improvements in the design of nuclear power plants make them safer than ever. Perhaps most importantly, nuclear power is the best hope for developing nations such as India and China. These and other countries are attempting to find a place for their large populations among the developed nations of the world. To do so, they need energy.

This point of view is not shared by all people. Many environmentalists believe that nuclear power plants are a disaster waiting to happen. Their views are sometimes supported by the mass media, which tends to focus on bad news rather than good. A documentary prepared by the Public Broadcasting Service (PBS) is a case in point. The documentary was titled "Meltdown at Three Mile Island." This title is dramatic, but it is false. No "meltdown" occurred at Three Mile Island.

BARRIERS TO IMPLEMENTATION OR ACCEPTANCE

The popular culture adds to the climate of distrust and emotional debate surrounding nuclear energy. Movies routinely depict scientists as "mad," as people bent on making scientific discoveries no matter what effects those discoveries might have on the human community. Cable-television science fiction channels routinely run movies about creatures that have been mutated into killer beasts because of science, especially nuclear science. At best, the stereotype of the scientist is one of an unappealing, slightly eccentric person. In this climate, the mysteries of nuclear power become an easy target for people's fears and uncertainties about the future.

During the first years of the twenty-first century, nuclear energy development was very much on hold, particularly in Western nations such as those in North America and Europe, as well as Australia. Public sentiment in the West favors other alternatives, such as solar, wind, and hydrogen. Less developed nations, though, do not have the luxury of picking and choosing, and many are going ahead with plans for nuclear power plants.

For More Information

Books

Angelo, Joseph A. *Nuclear Technology.* Westport, CT: Greenwood Press, 2004.

Domenici, Peter V. *A Brighter Tomorrow : Fulfilling the Promise of Nuclear Energy.* Lanham, MD: Rowman and Littlefield, 2004.

Heaberlin, Scott W. *A Case for Nuclear-Generated Electricity: (Or Why I Think Nuclear Power Is Cool and Why It Is Important That You Think So Too).* Columbus, OH: Battelle Press, 2003.

Kaku, Michio, and Jennifer Trainer, eds. *Nuclear Power: Both Sides.* New York: Norton, 1983.

Lusted, Marcia, and Greg Lusted. *A Nuclear Power Plant.* San Diego, CA: Lucent Books, 2004.

Morris, Robert C. *The Environmental Case for Nuclear Power.* St. Paul, MN: Paragon House, 2000.

Seaborg, Glenn T. *Peaceful Uses of Nuclear Energy.* Honolulu, HI: University Press of the Pacific, 2005.

Web sites

''The Discovery of Fission.'' Center for History of Physics. http://www.aip.org/history/mod/fission/fission1/01.html (accessed on December 17, 2005).

''Nuclear Terrorism—How to Prevent It.'' Nuclear Control Institute. http://www.nci.org/nuketerror.htm (accessed on December 17, 2005).

''Safety of Nuclear Power.'' Uranium Information Centre, Ltd. http://www.uic.com.au/nip14.htm (accessed on December 17, 2005).

''What Is Uranium? How Does It Work?'' World Nuclear Association. http://www.world-nuclear.org/education/uran.htm (accessed on December 17, 2005).

Solar Energy

INTRODUCTION: WHAT IS SOLAR ENERGY?

Solar energy is energy made from sunlight. Light from the sun may be used to make electricity, to provide heating and cooling for buildings, and to heat water. Solar energy has been used for thousands of years in other ways as well.

Most life on Earth could not exist without the sun. Most plants produce their food via a chemical process called photosynthesis that begins with sunlight. Many animals include plants as part of their diet, making solar energy an indirect source of food for them. People can eat both plants and animals in a food chain providing one example of the importance of the sun's energy.

In direct or indirect fashion, the sun is responsible for nearly all the energy sources to be found on Earth. All the coal, oil, and natural gas were produced by decaying plants millions of years ago. In other words, the primary fossil fuels used today are really stored solar energy.

The heat from the sun also drives the wind, which is another renewable source of energy. Wind arises because Earth's atmosphere is heated unevenly by the sun. The only power sources that do not come from the sun's heat are the heat produced by radioactive decay at Earth's core; ocean tides, which are influenced by the moon's gravitational force; and nuclear fusion and fission.

Historical overview: Notable discoveries and the people who made them

Ancient peoples did not just use solar energy; many of them worshipped gods based on the sun. More than 5,000 years ago ancient Egyptians worshipped a sun god named Ra as the first ruler of Egypt. Two ancient Greek gods, Apollo and Helios, were

Words to Know

Attenuator A device that reduces the strength of an energy wave, such as sunlight.

Convection The circulation movement of a substance resulting from areas of different temperatures and/or densities.

Current The flow of electricity.

Distillation A process of separating or purifying a liquid by boiling the substance and then condensing the product.

Heliostat A mirror that reflects the sun in a constant direction.

Hybridized The bringing together of two different types of technology.

Modular An object which can be easily arranged, rearranged, replaced, or interchanged with similar objects.

Passive A device that does not use a source of energy.

likewise identified with the sun. Shamash was a sun god worshipped in Mesopotamia.

Ancient uses of solar energy

Since at least the time when these gods were worshipped, the rays of the sun were used to dry things such as clothes, crops, and food. For centuries people who lived in the desert made homes from adobe, a type of brick made from sun-dried earth and straw. Adobe stores and absorbs the sun's heat during the day, which keeps the home cool. Then it releases heat at night to warm the home.

Ancient Greeks were aware of an early form of passive solar heating and cooling for homes. Passive solar heating and cooling use the sun's energy without help from any machines or devices. In one of his works, the philosopher Socrates (470–399 BCE [before the common era]) described how a home should be placed in relation to the sun so that it would be warmed in the winter and cooled in the summer. Ancient Romans and Chinese also designed and placed homes based on the principles of passive solar heating and cooling.

One famous Roman, Pliny the Younger (c. 61–c. 112), built a home in northern Italy that used this concept. In one room, he placed thin sheets of transparent mica (a mineral) in the window opening. That room was kept warmer than the others in the home. Because of the position of his house, Pliny was able to use less wood, which was used for heat and was in short supply.

Another way that ancient Romans used the principles behind passive solar energy was in the heating of water. In the public baths that were common at the time, black tiles were used in designs on the floors and walls. These tiles were set so they would be heated

Polar Bears and Solar Energy

Scientists have discovered that the fur and skin of polar bears are very effective at converting sunshine into heat energy. Researchers became interested in learning more about this effect when Canadian scientists found that polar bears could not be seen through infrared photography equipment. Infrared cameras are supposed to be able to detect anything that gives off heat, including all warm-blooded animals. But such cameras cannot see polar bears because their fur keeps the body heat inside so well that it cannot be detected on the outside of their bodies. A polar bear's white fur even converts more than 95 percent of the sun's ultraviolet rays into heat. This amount is larger than any solar technology that scientists and researchers have devised (come up with).

A polar bear's white fur converts more than 95 percent of the sun's ultraviolet rays into heat. *JLM Visuals. Reproduced by permission.*

Scientists have studied polar bear fur to determine why it is so efficient at drawing in and holding heat. There are several reasons why they think the fur works this way. Each piece of hair in polar bear fur is really not white, but transparent or clear. And each hair is hollow at its inner core. Because each hair is hollow, the light that hits the fur travels from the hair's tip to the skin of the polar bear. Though polar bear fur is white, the skin is black. So when the sunlight reaches the skin, it is converted into heat. Some researchers believe that this is because the hairs work the way fiber optic cable works when it transmits telephone calls. The hairs send the heat from the sun down the hair to the skin of the polar bear, like fiber optic cables transmit light from one point to another. However, other researchers do not agree and are unsure of the process by which polar bears retain their heat so effectively.

Scientists have used their findings on polar bear fur to improve flat plate collectors, photovoltaic (PV) cells, and other solar technologies. They have applied it to reduce heat loss in flat plate collectors. They are hoping that other applications outside of solar energy might be possible.

by sunlight. The water that ran to the baths would pour over the tiles and become warmed. A Roman architect named Vitruvius (died c. 25 BCE) drew up plans for a bathhouse that used passive solar design to heat the building. He oriented the building so that it would be warmed by sunlight in the late afternoon, especially during the winter.

There are also ancient examples of concentrated solar power. In the ruins of Ninevah in ancient Assyria, burning glasses were found. Burning glasses are like magnifying lenses. They could be used to start a fire by concentrating light from the sun into a beam.

Modern solar developments

Solar energy has been used for scientific purposes for several centuries. One scientist, Joseph Priestly (1733–1804), used sunlight to accomplish his discovery and isolation of oxygen in the 1770s. He heated and broke down mercuric oxide using heat created by concentrated sunlight.

An early nineteenth-century development was the greenhouse. Greenhouses are essentially passive solar energy collectors that collect the sun's energy to help grow plants. They capture light energy and retain heat while holding in humidity, which is used to water the plants. Greenhouses make it possible to grow plants even in winter.

Significant discoveries that advanced the use and efficiency of solar technology occurred in the nineteenth and twentieth centuries: photovoltaic cells and solar collectors, dish systems and trough systems, and power towers.

Photovoltaic cells

The idea behind the photovoltaic cell was described by Alexandre-Edmond Becquerel in 1839. This scientist discovered the photovoltaic effect (also known as the photoelectric effect). He made his findings while conducting an experiment on an electrolytic cell. This cell was made of photosensitive materials and consisted of two metal electrodes placed in an electricity-conducting solution. When this cell was exposed to sunlight, an electric current was created.

Becquerel's experiments inspired other scientists to continue to work on the photovoltaic effect. Another discovery came in 1873 when Willoughby Smith (1828–1891) discovered the photoconductivity of the element selenium. Four years later two other scientists, William G. Adams and R. E. Day, learned that solid selenium could be used in the photovoltaic effect. They developed the first photovoltaic cell made with selenium. Their cell had limited power: It could convert less than 1 percent of the energy of the sun into electricity.

Though the photovoltaic cell designed by Adams and Day was not very powerful, another inventor was able to improve on their design. In 1883 the American scientist Charles Fritts came up with his own photovoltaic cell, which was made from selenium wafers. While work continued on photovoltaic cells in the late nineteenth and early twentieth centuries, it was not until 1954 that the first practical version of photovoltaic cells was created.

This cell was made in Bell Laboratories by three scientists: Calvin Fuller, Daryl Chapin, and Gerald Person. In the early 1950s they created a photovoltaic cell that was made from crystalline silicon. When exposed to light, their creation produced a significant amount of electricity. The 1954 version of the photovoltaic cell has proved to be the basis of all future photovoltaic cells. It was patented in 1957 and called a "Solar Energy Converting Apparatus." It has since been used on nearly all space satellites since that time.

The first satellite to use photovoltaic cells was the Vanguard 1, launched in 1958. The success of the Vanguard 1 led the National Aeronautics and Space Administration (NASA) to use photovoltaic cells as the normal way of powering satellites in the Earth's orbit. Even the Hubble Space Telescope, which was launched in 1990, uses photovoltaic cells to produce electric power. Such cells are also used to power the international space station.

Dish systems, trough systems, and power towers

In the mid-1800s a French engineer and math instructor named Auguste Mouchout was granted a patent for solar technology that used the sun to make steam. Mouchout used a dish to concentrate the sun's rays. His invention was an early version of the dish system. He began working on the project in 1860 in part because he was concerned that his country was too dependent on coal as an energy source.

Mouchout's design featured a cauldron filled with water. It was surrounded by a polished metal dish that focused the sunlight on the cauldron. This focused sunlight created steam that powered an engine. Mouchout's original engine generated one-half horsepower.

Over the next twenty years Mouchout continued to improve on his design. He replaced the cauldron with a multi-tubed boiler. This boiler made the engine run even better. Mouchout also made his overall design bigger. However, Mouchout's invention only found limited applications. It was used in the French protectorate of Algeria as a source of power for a time. Even this utilization was only short-lived, as coal transportation to Algeria improved and coal remained a much cheaper source of energy. Despite this situation,

Mouchout was well known in France in his time, had the backing of the French government, and won a medal for his work.

Mouchout's invention led to innovations on the dish system by other scientists. One of them was John Ericsson (1803–1889), an engineer who was a native of Sweden but who lived in the United States. In the 1870s Ericsson came up with a different version of Mouchout's means of using the sun to make power. Ericsson attempted to improve on Mouchout's design. He first replaced the dish with a reflector shaped like a combination of a cone and a dish.

Ericsson later replaced this conical dish shape with a parabolic trough. This trough looked like an oil drum cut in half lengthwise. The trough reflected the sun's radiation in a line across the open side of the reflector. What Ericsson came up with evolved into the trough system that is currently used to convert solar energy into electricity.

Ericsson's creation was simple to make. It tracked the sun in a single direction: either north to south or east to west. The trough could not produce the same temperatures or work as efficiently as the dish-shaped reflector. However, Ericsson's design was functional from the beginning. Until his death, he continued to try to improve his design with lighter materials for the reflector.

Another scientist worked with Mouchout's basic design to create a new technology that became important in the late twentieth century. In 1878 William Adams, an English scientist, came up with a solar technology design that would become the basis for power towers. Adams set up flat, silvered mirrors in a semicircle around a cauldron. The mirrors were erected this way so that sunlight could be continuously focused on the cauldron. The mirrors were also placed on a rack that moved along a semicircular track so they could be moved throughout the day around the boiler by an attendant. Most modern solar power towers also use mirrors placed in a semicircle that reflect sunlight onto a boiler that generates steam to run a heat engine. Adams was able to run a small engine with his invention, though it never moved beyond the experimentation stage.

The American scientist Aubrey Eneas worked with both dishes and troughs, as well as with other solar technologies, in the late nineteenth and early twentieth centuries. Eneas first began experimenting with solar-driven motors. He formed the first solar company, the Solar Motor Company, in 1900 and spent the next five years working on his idea. Eneas first made a reflector similar to Ericsson's, but he could not make it work.

Then Eneas focused on making a reflector more like Mouchout's. Eneas improved on Mouchout's design to make the dish larger by increasing the sides to be more upright. The dish focused the sunlight on a boiler that was 50 percent bigger than earlier versions. Eneas exhibited his design at a Pasadena, California, ostrich farm. His demonstration model had a 33-foot diameter reflector with 1,788 mirrors. The boiler could hold 100 gallons (378 liters) of water and was 13 feet (3.9 meters) long. While Eneas received some attention in the press and sold a few of his systems, none could withstand bad weather. His idea failed to catch on.

Solar collectors

In the 1880s a French engineer named Charles Tellier (1828–1913) made significant strides in the development of the solar collector. He designed the first nonreflecting (that is, nonconcentrating) solar motor. His work in this area led to research for which he was better known: refrigeration.

Tellier's solar collector was made up of ten plates. Each plate consisted of two iron sheets that were riveted (joined) together so they had a watertight seal. The plates were connected by tubes to form a solar collector. Inside the collector, Tellier placed ammonia instead of water because ammonia has a lower boiling point than water. In 1885 he put such solar collectors on the roof of his home. When the collector was exposed to the sun, each plate released ammonia gas.

Tellier's solar collector worked well. The pressurized ammonia gas powered a water pump. This water pump was put in a well and was able to pump about 300 gallons per hour during daytime hours. Tellier was able to increase the efficiency of his collectors by covering the top with glass and by putting insulation on the bottom.

Tellier believed that his solar collectors would work for anyone in the Northern Hemisphere that had a south-facing roof. He also was certain that his system could be used industrially if more plates were added to the collectors to make the system bigger. Tellier hoped his invention would be used in Africa to provide power and to manufacture ice. But while he realized that he had a good idea, Tellier decided to focus on developing refrigeration technologies.

Other inventors improved on Tellier's design. In the first decades of the twentieth century American scientists such as Henry Willsie and Frank Shuman came up with their own solar collector designs. Their inventions failed to catch on at the time but continued to improve the technology.

The Million Solar Roofs Initiative

Announced by the U.S. government in June 1997, the Million Solar Roofs Initiative called for one million homes and businesses in the United States to install solar energy technologies such as PV cells for electricity, solar collectors, and solar water heaters by 2010. The initiative had several goals. The federal government hoped to increase the market for solar energy and keep it viable. It was also hoped to spur job creation in the solar industry in the United States. One study showed that each solar roof could stop thirty-four tons of greenhouse gases from reaching the atmosphere over its lifetime of use. There was widespread support for the initiative. At least eighty-nine different partnerships formed to help achieve this goal, with both state and local governments as well as private businesses and community organizations. Financial incentives were given by the U.S. Department of Energy and by agencies on the state and local levels. By 2002 nearly 350,000 roofs had been installed as part of the program.

Government-supported developments

Government support of solar energy helped move the industry forward in the 1970s and early 1980s. Many homes were built that featured solar technologies. Although government support decreased in the 1980s and early 1990s, some progress continued on alternative energy research. By the mid-1990s there was renewed interest in the United States in building homes and businesses that used solar technologies.

In 2004 only six percent of U.S. energy came from renewable sources, and only three percent of that six percent came from solar energy. However, many experts believe that solar power will be the most important alternative energy source in the future.

How solar energy works

Solar energy technologies use the energy that comes from the sun. Inside the sun, hydrogen atoms combine to make helium, and the process produces the extreme amount of heat that is felt on Earth. The core of the sun has a temperature of 36,000,000°F (20,000,000°C). The surface of the sun, called the photosphere, has a temperature of 10,000°F (5,538°C). The energy that the sun creates has to travel 93,000,000 miles (150,000,000 kilometers) to reach the surface of Earth.

People on Earth do not feel the full force of the sun, because Earth's upper atmosphere blocks out much of the sun's thermal power. This power, sometimes called radiation, is spread out when it hits the water vapor, molecules of gas, and clouds that surround Earth. The sunlight that does reach the ground is called direct radiation or beam radiation. If the sunlight hits something before reaching the ground, it is called diffuse radiation.

The amount of solar radiation that reaches the surface of Earth is more than ten thousand times the amount of energy used by the world already. A significant amount of the sun's radiant energy, about 69 percent, is reflected back into space by such things as clouds, ice found on the ice caps, land, and bodies of water. Of the energy that is absorbed by Earth, about 70 percent of the absorption is done by the oceans. Solar energy helps keep the oceans from freezing and pushes their currents. It also prevents Earth's atmosphere from freezing.

Current solar technology

Solar technologies can be divided into passive systems or active systems. Passive solar energy projects only employ the sunlight; no other forms of energy are used. Active solar energy systems employ additional mechanisms such as pumps, blowers, or generators to apply or add to the solar energy created. Active systems often make electricity or heat. Solar water heating systems can be either active or passive.

Passive solar systems

Passive solar systems are primarily concerned with the design of buildings, homes, and lighting. Passive solar design focuses on the placement of the home or building and on windows, ventilation, and insulation to cut down on the need for electricity by using the sun. The home or building is designed to maximize the potential of solar energy for heating and cooling. In northern countries such as Canada, where sunshine is not as strong as it is in locations to the south, passive solar heating is one of the easiest forms of solar technology to use.

One important form of passive solar design is known as "daylighting." In daylighting the placement and design of windows is used to encourage natural sunlight to light the inside of a building instead of electric lights. Daylighting helps cut down on lighting costs, and many experts believe that exposure to natural rather than artificial light sources provides health benefits to humans.

Another type of passive solar system is the transpired solar collector. This is a relatively new passive solar technology made of dark perforated metal. Transpired solar collectors are used to heat buildings by heating the air. They can also cool buildings in summertime.

Active solar systems

Active systems include solar collectors (also known as solar panels), which are primarily used on solar hot water heaters; photovoltaic (PV) cells, which make electricity; and concentrated solar power systems (also known as solar thermal systems), which also make electricity but on a larger scale than PV cells.

Solar collectors are used primarily to capture solar energy for use in solar hot water heaters. However, they can also be used to provide heat in a building and even to make the energy to cool a building. While not all solar collectors are used in active solar energy systems, it is more common for solar collectors to be used in an active system than a passive system.

Photovoltaic (PV) cells convert sunlight directly into electricity inside the cell. They are more adaptable than many other types of solar energy technology. In addition to powering satellites, PV cells can be put on buildings to provide electricity for any number of uses. They do not require direct sun to convert sunlight into electricity.

There are at least five types of concentrated solar power systems that focus the sun's power to make electricity on a larger scale than PV cells. They include solar ponds, parabolic trough systems, dish systems and dish-engine systems, solar power towers, and solar furnaces. Mirrors or other reflective devices draw in as much sunlight as possible to these systems. They often track the sun as it moves through the sky in order to capture the most sunlight.

Concentrated solar power systems usually heat water, or another fluid that is connected to a source of water, to make steam. The steam is used to drive turbines that create electricity. Concentrated solar power systems are primarily used for industrial applications and to make electricity for consumers and businesses on a wide scale.

Emerging solar technologies

There are several technologies being developed that bypass mirrors and collectors to capture the sun. Solar paints contain conductive polymers, extremely small semiconducting wires, or quantum dots. Such paints could be used to coat any surface and turn it into an electrical generator. Other companies are working on similar technologies for plastics. Rolls of plastic are coated with an electricity-generating film. The plastic could be spread over roofs or other surfaces to convert sunlight into electricity.

The use of solar energy to cool homes and buildings is another area under more development. Such systems use solar panels to produce electricity. These panels power a pump connected to an absorber machine. This machine works something like a refrigerator. The absorber employs hot air to compress a gas. When this gas expands, it causes a reaction that cools the air. Solar thermal coolers are expected to reach the commercial market in the early twenty-first century.

Many new solar technologies are still in the experimental stage. One possibility is solar-powered air flights. Another is a different kind of solar lighting, in which a building's interior is lit by a parabolic collector on the roof. This collector is connected to the interior by fiber optic light pipes. Such a system would make its own electricity to power the lights.

Benefits and drawbacks to solar energy

One of the primary benefits to solar energy is that it is a renewable resource. Sunshine is available everywhere free. There is no limit to its renewability, at least not until the sun burns itself out billions of years from now. Solar energy also does not contribute to pollution and thus is considered a "clean" energy source. Using it produces no greenhouse gases and thus does not contribute to global warming.

The biggest drawback to using solar energy is the cost of the technology. Solar photovoltaic cells and solar collectors are still very expensive. While the technology may become cheaper over time, it is still costly when compared to the amount of energy it will produce over its use cycle. Similarly, it is very expensive to build solar power towers and furnaces. Using such technology to generate power on a wide scale is too expensive to be used realistically, at least as of the early twenty-first century.

Another major problem with solar technology is that solar energy is not available on demand in every location on Earth. Heavy cloud cover can limit the use of some solar energy systems. Some systems cannot be used at all if direct sunlight is not available. In most areas of the world, only low-power solar energy applications can be used because of the lack of direct sunlight.

Japan and Germany Lead the Way

No two nations have invested more heavily in solar power than Japan and Germany. By 2001 Japan was able to produce up to 671 megawatts of solar-generated power at peak conditions. The country was also a leader in the number of solar water heating units being used. As of 2005 there were more solar hot water heaters being used just in the city of Tokyo than in the whole of the United States.

As of the early 2000s Germany was number two in the world with 260.6 megawatts of solar-generated power being produced at peak conditions. By this point the German city of Freiberg had more solar projects than any other city on the continent of Europe. It was home to the headquarters of the International Solar Energy society, and the city also featured parking meters powered by solar power.

For large-scale projects such as solar power towers and solar furnaces, or even smaller-scale projects such as solar ponds, dish systems, and trough systems, large areas of land are needed. In the desert, where a number of these systems are currently located, the solar technology that is put there to capture the intense sunshine is considered unsightly by some people.

Environmental impact of solar energy

Solar energy can have both positive and negative effects on the environment. On the positive side, most solar technologies are environmentally friendly. They do not pollute the atmosphere by emitting (giving off) greenhouse gases, they do not produce radioactive waste like nuclear energy reactors, and they do not contribute to global warming or acid rain. Most solar energy systems are silent or quiet when they operate, which cuts down on noise pollution. If solar technologies that make electricity on a significant scale can be adopted, many countries can lessen their dependence on electricity produced by fossil fuels. This change could decrease the amount of environmental pollution in the world.

However, solar energy technologies are not perfect. In addition to large-scale projects negatively affecting the landscape, these solar technologies can negatively affect the animal life around them. Big dish systems, trough systems, and power towers take up land that animals live on and affect their habitats. The very building of these projects can pollute otherwise pristine (clean) lands, even if the solar technology itself does not. Also, while the use of solar technology does not pollute the environment, the manufacture of certain types of solar technology can.

Economic impact of solar energy

The adoption of solar energy technologies can have a profound impact on the economies of individual communities, states, and countries. When renewable energy sources such as solar energy are used in a community in the United States, more of the money spent on that energy stays at least in the same area, if not within the country. Most of the cost of solar energy implementation comes from materials and installation, not buying the actual fuel source as is the case with oil. The materials can be local, and the installation is often done by local companies.

The use of solar energy can also make countries more energy independent. Currently many countries rely on foreign oil for nearly all their energy needs. Because a few countries hold most of the oil resources in the world, they have a lot of control over the pricing and distribution of that oil. If nations are able to augment the imported oil with solar energy, they will be better able to govern their future energy supply.

Societal impact of solar energy

The spread of solar energy technologies could lead to electrical power being available where it was not available before. People who live in rural areas are often not connected to an electrical power grid; this is especially true in poorer, less developed countries. In 2000 more than two billion people worldwide did not have access to electricity. Solar technologies could provide energy to these communities.

Barriers to implementation or acceptance

There are two main barriers to implementation of solar energy on a larger scale: efficiency of the technology and cost of the technology. As of 2005 the existing solar technology was still too inefficient to make it a viable energy source on a large scale. The existing PV cells, for instance, do not convert enough sunlight into energy.

The other main barrier is cost. Over the years many researchers and companies have announced that solar technologies will be ready and/or profitable by a certain date, but this promise has not been kept. Even if the technology has become available, it has not been developed as cheaply as promised. Some critics believe that solar energy, as well as other alternative energy sources, will never live up to the promises made by its supporters; they feel that the energy produced by solar power will never be enough to make up for the high cost of producing it. Increased tax breaks for solar technology on the federal, state, and local levels could help build the marketplace for the technology and drive down the production and implementation cost.

Another barrier to implementation is that solar technology has not yet been applied on a widespread basis and thus remains unproven on a large scale. The technology has done well in small, specialty markets, proving that it can work, at least on this scale. More large-scale success would increase the perception of solar energy as a useful technology for the future.

PASSIVE SOLAR DESIGN

Passive solar design focuses on the construction of the building, the way its site is set up, the environment around it, and its orientation to the sun to make the best use of the amount of sunlight to which it is exposed. These choices can cut down on electricity costs for the building while also helping to light, heat, and cool it.

Passive solar design can be used on many types of buildings, including homes, businesses, industrial sites, schools, and shopping facilities. In the Northern Hemisphere, buildings created on the principles of passive solar design usually have the longest walls running from east to west. This orientation allows heating from the sun in the winter and much less sun exposure in the summer. Such buildings also feature large south-facing windows, which are often insulated. Building materials that absorb and slowly release the heat of the sun are used in the flooring and walls. Such building materials include rocks, stone, or concrete; some even contain saltwater, which can collect the solar energy as heat.

Another key facet of passive solar building design is a roof overhang. Such overhangs are designed to allow sunlight to stream inside during the winter and shade windows from the higher sun in the summer. In areas where summer temperatures are high, especially in the South, putting roof overhangs on buildings can help keep buildings much cooler than they otherwise would be.

Some passive solar-designed buildings can be located underground or built into the side of a hill. Because the temperatures found a few feet below ground are steady, this allows the building

FOR SALE
AWARD WINNING
PASSIVE SOLAR HOUSE
George Guttmann - Contractor
Van Horne & Van Horne - Architects
For Sales Information:
322-8940
Susan Moss 322-6418
Realtors

Passive solar design focuses on the placement of the home or building and on windows, ventilation, and insulation to cut down on the need for electricity by using the sun. © *Joel W. Rogers/ Corbis.*

to be cool in the summer and warm in the winter. Another passive solar concept is landscaping, or the design and placement of trees and shrubs around a building. For example, deciduous trees, which lose their leaves in the winter, can be planted around the building to keep it cool during the summer by providing shade. During the winter, when the trees are bare, more sunlight reaches the building.

There are five basic types of passive solar design systems:

1. Direct Gain. Direct gain is the simplest type of passive solar design. In this system a large number of windows in a building are set up to face south (in the Northern Hemisphere). The glass is usually double-paned or even triple-paned. That is, the glass consists of two to three panes of glass with a pocket of air in between each pane. These panes are sealed inside one frame. Materials that can absorb and store the sun's heat can be incorporated into the floors and walls that are hit by the sun. These floors and walls release the heat at night, when it is needed the most to heat the building.

Adobe house with passive solar power. © *Michael Freeman/Corbis.*

2. Thermal Storage. Thermal storage is very similar to direct gain. In this system, there is also a large wall oriented to the south in the Northern Hemisphere. This wall is placed behind double-glazed windows so that it can absorb sunlight. In some of these thermal storage systems, the wall contains a storage medium such as masonry or perhaps water. The solar energy that is collected is stored during daylight hours so that it can be released when there is no sun.
3. Solar Greenhouse. Solar greenhouses are also known as sunspaces. They are a combination of both direct gain and thermal storage but are located in a greenhouse. The wall of the thermal storage system is placed next to the greenhouse and the home to which it is attached. This system primarily heats the greenhouse but also can provide heat to the house itself.
4. Roof Pond. As its name implies, the roof pond system consists of ponds of water placed on a roof. These ponds, which are exposed to the sun, collect the radiation from the sun and store it. The heat that is produced is controlled by

insulating panels that are movable. During the winter these panels are open during daylight hours so that sunlight can be collected. During nighttime hours the panels are closed so that little or no heat is lost. The heat that is collected is released into the building to warm it. During the summer roof ponds are used in the opposite way. The panels are closed during the day to block the heat of the sun. At night they are opened to allow cooling of the building.

5. Convective Loop. The convective loop is also known as a natural convective loop. In this system, a collector is located below the building's living space. The hot air that is created from solar energy rises to heat this living space when needed.

Current uses of passive solar design

Passive solar design is primarily used in the planning of homes, offices, schools, and any other type of building. In 2001 about one million U.S. homes and twenty thousand buildings used only for commercial purposes employed the principles of passive solar design.

Benefits and drawbacks of passive solar design

What makes passive solar design so simple is that it has no moving parts or working parts. Buildings made using passive solar design do not need to be maintained any differently than any other type of building.

Buildings created with passive solar design in mind are more effective in sunny environments, though buildings in any environment benefit from passive solar design. Sometimes these buildings can become overheated in the summer. However, design changes can address this issue. Nevertheless, it would be difficult to retrofit a home or building with passive solar design principles unless it sat on its lot in the correct orientation to the sun.

Impact of passive solar design

Passive solar design has no real negative effects on the environment, other than what would happen when any building is constructed. The principles of passive solar design often incorporate trees, resulting in more trees being planted in an area.

Economically, passive solar–designed buildings can produce heating bills that are 50 percent less than buildings without any passive solar design principles, a significant savings in energy costs. The increased use of passive solar design can bring business to builders specializing in this discipline. However, unlike other solar technologies, passive solar design does not afford any tax breaks from the U.S. government.

Issues, challenges, and obstacles of passive solar design

One potential issue related to the use of passive solar design is that not every architect accepts and employs these principles. There are only a limited number of professionals who design such buildings. There is currently a limited market for passive solar design because many people do not know about it. However, the popularity of passive solar design is poised to grow as consumers look for ways to battle higher heating and cooling bills caused by the increase in the cost of electricity and natural gas.

DAYLIGHTING

Daylighting, also known as passive lighting, is a form of passive solar design. Daylighting involves the use of sunlight to light up the inside of a building. Daylighting can fully replace electric lights, or it can be used to cut down on electrical costs by supplementing electrical lighting already being used. Daylighting can also be used to heat a building.

Daylighting primarily occurs through a building's windows, though other kinds of openings on buildings, such as skylights, can also be used. The windows are often large and, in the Northern Hemisphere, face south. Buildings and homes that use daylighting have specific placement and spacing of windows. For example, windows that are higher up on a wall distribute sunlight better. Windows called clerestory windows (a row of windows located at the top of a wall, near the roof) are an important part of daylighting in museums and churches. Skylights, when combined with sensors and other lighting elements, can ensure that lighting inside a building stays even.

Windows used in daylighting absorb sunlight and release it slowly to light up a building. One way to regulate the amount of sunlight and/or heat is through window shades or curtains that are insulating. Light shelves can also be used. They are placed so that the sunlight drawn in by the windows is reflected and lights a room from top to bottom. These shelves can bring natural light deeper into a room.

Chemical compounds in windows for daylighting can be made part of window glass or placed between the panes of double- and triple-paned windows. These compounds can boost how much solar energy a window can store. They can also increase the insulating capacity of windows. In addition, coatings and glazings on the windows can control the amount of light or heat. The heating effect of daylighting can be increased by window coatings that are anti-reflective. Some window coatings can carry an electric current that can moderate how much light or heat is let in based on current weather conditions. One type of glazing can allow a measured amount of light to pass through a window while keeping heat out.

In daylighting systems where natural light is used with electrical lighting, there is need for a control system. This control system regulates the amount of electric light used based on how much daylighting is available. The types of controls include photocell sensors, infrared receivers, occupant sensors, dimming control systems, and wall-station controls.

Building materials and interior design can enhance the effectiveness of daylighting. Walls that are white or brightly colored reflect the light that is drawn inside. In office buildings, cubicle walls kept under a certain height will allow the sunlight to spread over the office.

New technologies are being developed to increase the effect of daylighting. Some buildings are incorporating heliostats, which are the same mirrors used in solar power towers. The heliostats can track the movement of the sun during the day and reflect the sunshine into windows. Another device that is being worked on employs fiber optics to take the sunlight collected on the roof inside the building.

Benefits and drawbacks of daylighting

As daylighting provides light during the day, the amount of heat gain from electric lighting is reduced significantly. Daylighting also makes homes and buildings less gloomy. However, homes and buildings that use daylighting often have to deal with issues such as heat and glare. If the natural lighting is not regulated, the system is not properly designed, or the correct type of window for the local environment is not used, homes and buildings can become hotter than they would were daylighting not used. Daylighting can potentially increase cooling costs during the summer because there is more natural light inside. Daylighting will not work everywhere because there is not enough sunshine in some locations.

Daylighting is difficult to incorporate into buildings that have already been constructed. Even if daylighting is built into a new building, the controls needed to regulate the natural light and electric lights are expensive and require a significant investment. After the system is installed, it must be operated and maintained. People must be trained to deal with the sensors and computer systems that come with many daylighting systems.

Impact of daylighting

There is little to no negative environmental impact with daylighting as an energy system. The only effect on the environment comes from the production of the windows, coatings, controlling systems, and buildings. Using daylighting ensures that fewer fossil fuels are burned, cutting down on pollution.

For consumers and businesses, the use of daylighting can cut electric bills significantly, perhaps up to one-half. It can also cut down on energy costs for buildings. If daylighting is done correctly, less air conditioning is needed during the summer months.

Because of daylighting's positive effects on people, workers in offices with daylighting are more productive. There are fewer absences and errors by such workers. When workers' productivity is increased, businesses can become more successful. Daylighting can even affect shoppers. Shopping centers and malls that incorporate daylighting into their design find that more natural light may lead to increased sales.

Issues, challenges, and obstacles of daylighting

Though daylighting is simple and the principles behind it show evidence of success, there is still a reluctance to embrace this solar energy system. The reasons vary. Adjusting building plans in order to place windows to save on electrical costs may increase the price of the building and thus affect its appeal to potential buyers. Also, daylighting is difficult to incorporate into existing buildings, so its growth may be limited solely to the new construction industry.

TRANSPIRED SOLAR COLLECTORS

A transpired solar collector, sometimes known under the brand name Solarwall, is used to heat what will become ventilated air as it enters a building. This relatively new technology was developed with the support of the U.S. Department of Energy and has won several awards.

The transpired solar collector is very simple. It is a metal panel that is dark colored and has perforations (lines of holes). The metal is usually corrugated steel or aluminum. The piece of metal is formed to fit and mounted on the outside of a south-facing building wall. The collector is not fully attached to the inside wall; instead, a gap is left between the metal panel and the interior wall of the building. There are ventilation fans at the top of the space and the interior wall. These fans draw in the air through the holes in the metal panel. After the air enters the space between the walls, it rises to the top of the panel. The air becomes heated as it passes near the hot metal panel and continues to rise to the ventilation fans, where it is sucked into the building. This hot air is circulated through the building via its air ducts.

A transpired solar collector does not just heat the air for a building. It can help cool the building as well. During summer months the ventilation fans draw in the hot air. Instead of bringing this hot air into the building, bypass dampers are used to move the hot air back outside. This hot air then does not come in direct contact with the inner wall, thus making the building cooler.

Current uses of transpired solar collectors

Transpired solar collectors are primarily used to heat air for office buildings, schools, homes, and industrial facilities. While the technology can be used in most buildings, it is really useful for buildings that are used by industry, commercial interests, and institutional interests. Such buildings usually need a lot of ventilation, and this technology can be extremely helpful in such circumstances. Transpired solar collectors can be used to preheat combustion air for industrial furnaces. In an agricultural setting this technology can be used to create hot air for crop drying.

Benefits and drawbacks of transpired solar collectors

As a means of heating air, transpired solar collectors are very inexpensive to make and very efficient. They preheat air twice as effectively as any other type of solar heater. Transpired solar collectors can use as much as 80 percent of the solar energy that comes into contact with the collector. The use of a transpired solar collector can result in much lower energy costs for the building to which it is attached.

Transpired solar collectors can be used in parts of the world where there is not a significant amount of direct sunlight. For example, this solar technology can be used in Canada and the northern United States. Snowfall can actually make the transpired solar collector heat better. When snow covers the ground, it can reflect as much as 70 percent more solar radiation onto the transpired solar collector. More reflected solar radiation results in more heat produced. In addition, transpired solar collectors do not need as much additional heating as other solar heating systems when there is no sunlight. The heat that is collected during the day can be retained and used after dark.

On the other hand, only buildings that have a south-facing wall, at least in the Northern Hemisphere, can effectively use a transpired solar collector. Because of this requirement, it can be difficult to retrofit certain homes and buildings with this solar technology.

Impact of transpired solar collectors

The use of transpired solar collectors has no real negative environmental impact. There is a chance that the manufacture of the metal or other pieces needed for the collector can negatively affect the environment. But by using a transpired solar collector, fossil fuel use can be lessened because the solar technology cuts down on energy costs.

Many states offer consumers and businesses tax credits and incentives for the installation and use of transpired solar collectors. In new construction projects, when the transpired solar collector begins to operate to both heat and cool homes and buildings, consumers and businesses save money. The technology can reduce annual heating costs by about two to eight dollars per square foot because it can increase the temperature of incoming air by 54°F (12°C). For new construction, transpired solar collectors can pay for themselves in three years. If the technology is put on an already existing building, the transpired solar collectors pay for themselves in seven years. The cost savings depends on how long the heating season is and what kind of air ventilation is needed.

Issues, challenges, and obstacles of transpired solar collectors

Transpired solar collectors have not yet been widely embraced because the technology is relatively new. The collectors were not invented until the 1990s, and the general public only has minimal knowledge of the technology.

Another obstacle is that transpired solar collectors are most often large and very noticeable on a building. Because they need a dark color, they do not always blend in with their surroundings. Certain types of businesses might be reluctant to put something so large on their building if the owners or operators feel the collector will detract from the way their building looks.

SOLAR WATER HEATING SYSTEMS

A solar water heating system uses the sun's power to heat water. The water can be used in homes, businesses, swimming pools, hot tubs, and spas. On a larger scale, water can be heated for industrial processes.

While there are many different types of solar water heating systems, there is a common method to how they work. Most are simple in design and inexpensive to install, even in older homes. In general, the sunlight passes through a collector. The radiation that is absorbed by the collector is usually converted to heat in a liquid-transfer medium or through the air. The radiation can also be used to heat the water directly.

Solar collectors are used primarily to capture solar energy for use in solar hot water heaters. However, they can also be used to provide heat in a building and even to make the energy to cool a building. © *Dietrich Rose/zefa/Corbis.*

Solar water heating systems can be active or passive to transfer the heat. An active solar water heating system uses pumps to transfer heat from the collector to the storage tank. Active systems can use a PV module to produce the electricity to run an electric pump motor. In a passive system, the system does not use pumps or control mechanisms to transfer the heat created to the storage tank. Instead, passive systems use natural forces such as gravity to circulate the water. There is also an exchange/storage tank of some kind. When such systems are used for bigger buildings that house businesses or offices, there is often more than one storage tank for the water.

There are at least six types of solar water heating systems:

1. Direct Systems. Direct systems use a pump to circulate the water. The water moves from the home into a water storage tank and passes through the solar collectors for heating.

After it leaves the collector, the water returns to a tank. From there, it is pumped back into the house as hot water. The pump can be powered by a PV cell or by an electronic controller or appliance timer. Direct systems are usually used in warm climates with few or no days in which the temperature dips below freezing. Because of this requirement, there is a very limited area where direct systems can be used, at least in the United States.

2. Indirect Systems. Indirect systems use a heat exchanger that is separate from the solar collector. The collector contains an antifreeze solution instead of the water to be heated. The heat exchanger transfers the heat from the collector's antifreeze solution to the water located in the water storage tank. The heat exchanger can either be inside the storage tank or outside the storage tank. One advantage to this system is that it can be used in areas where the temperature falls below the freezing point.
3. Thermosyphons. A thermosyphon solar water heating system features an insulated storage tank that is placed above the solar collector, usually a flat-plate collector. When the sun hits the collector, it warms the water located in the tubes that pass through the collector. This water travels up through the top of the storage tank, which is insulated, and out through a hot water pipe. At the bottom of the storage tank is the cold water, which travels down through a pipe and into the collector. Sometimes, a small pump can be added to this system if it is not possible to place the tank on the same level or below the collectors. This system is more common outside of the United States and can only be used in warmer climates where temperatures remain above freezing. Locations in the Caribbean, Middle East, Mediterranean, Australia, and Asia use this system.
4. Draindown/Drainback Systems. Draindown systems are often used in cold climates. In this system, water passes through the collector to be heated. Draindown systems prevent water from freezing inside the collector by the use of electric valves. These valves automatically remove the water from the collector if the temperature gets too cold. The drainback system is very similar to the draindown system. When the circulating pump that is part of the drainback system stops as a result of cold temperatures, the collector is automatically drained.

5. Integral Collector Storage (ICS) Systems. These types of systems are also known as integrated collector systems, batch heaters, bulk storage systems, or breadbox heaters. Whatever the name, the ICS system features a collector and 40-gallon (151-liter) insulated storage tank that are part of one unit. The tank is lined inside with glass and painted black to draw in the sun's heat. The ICS system is usually placed on a roof or in a place on the ground where there is sunlight. Cold water comes into the ICS system from the plumbing in the house. The inlet inside the tank pushes the water to the bottom of the tank. The hot water rises in the tank and goes into the building through an outlet. There can also be a backup tank below the ICS unit that transfers water to be heated when the already heated water is taken from the primary storage tank. One drawback to this system is that the hot water created by the ICS system should be used during the afternoon or evening hours. If it is not, it should be transferred into another storage tank before nightfall. Otherwise, the water in the primary storage tank might lose much of its heat overnight, especially in cold weather.
6. Swimming Pool Systems. The solar energy systems used to heat swimming pools and hot tubs are usually simpler than other kinds of solar water heaters, but just as effective. The use of a solar water heater can allow an outdoor pool or hot tub to be used for at least four months longer than a pool or hot tub without a heater. The system usually consists only of a temperature sensor, an electronic controller, a pumping system, and solar collectors. The collectors can be mounted on the pool's deck, on the ground, or on a roof. Most collectors used for pools or hot tubs usually have no glass covering or insulation. They are also usually lower-temperature collectors. That is, they usually are designed only to raise the temperature of the pool's water to about 80 to 100°F (26 to 37°C). This system does not need a storage tank since the pool or hot tub serves as the storage medium.

There are also pumped systems intended for bigger buildings, such as hotels and gymnasiums. In this type of system the storage tank is located inside the building and uses a pump to transfer water between the collectors and the tank. In addition, a controller is needed that detects when the water in the panels is hotter than the water in the tanks. The controller regulates the pump so that the temperatures remain correct. If the outside temperature gets below freezing, the pump starts running to prevent the water from freezing.

Flat plate collectors

The most common type of energy collector, the flat plate collector, is a rectangular-shaped box that is put on the roof of the home or building where the solar water heating system is located. Inside the box is a thin absorber sheet, usually black in color and made of either copper or aluminum. Behind the sheet is a tubing system in the form of a grid or coils. The collector and tubing system are put inside an insulated casing. The cover is usually glass and transparent. This glass is often black or a dark color that draws in the sunlight.

As the sun shines, the heat builds up in the collector and heats the fluid that is inside the tubes. If it is water, it is heated and passes through a storage tank. If the fluid inside is antifreeze, the water is heated by circulating the heated solution through a tube inside the storage tank in which the water is located.

Evacuated tube collectors

This type of collector features rows of glass tubes placed parallel to each other with a vacuum between them that insulates the tubes and helps hold on to the heat. The tubes are also transparent and covered with a coating. Inside each tube is an absorber with liquid inside it. When light from the sun hits the tube and its radiation is absorbed by the absorber, the liquid inside is heated. Because of the vacuum between the tubes, this liquid can be heated to very high temperatures, up to 350°F (176°C). Though the evacuated tube collectors can achieve high temperatures, they are more fragile than other types of solar collectors and more expensive.

Current use of solar water heating systems

Solar water heating systems have existed for many years. They are used in homes, businesses, schools, office buildings, prisons, military bases, and industrial settings. Solar water heating systems can be used to power irrigation systems, and they can also be used to provide water for livestock on farms and ranches. Solar hot water heating systems are often used where natural gas or electricity cannot be used to heat water.

For a typical household, solar water heating systems can provide from 70 to 90 percent of the hot water needed for bathing and laundry. In a common single family home in the United States, about

25 percent of the energy is used to heat water. As of 2001 about 1.5 million solar water heating systems were being used in the United States in both commercial businesses and homes. About 300,000 swimming pools were being heated the same way. By 2005 at least 500,000 homes in California alone used solar water heating systems.

Benefits and drawbacks of solar water heating systems

One of the biggest benefits to solar water heating systems is their practicality. They are relatively easy to install in both new and existing homes and buildings. Because of the variety of systems available, at least one type will work in most locations. The systems are long-lasting, with most systems lasting a minimum of fifteen to twenty years. Passive solar water heating systems in particular are very inexpensive because of the limited equipment involved and the little maintenance required.

However, certain types of solar water heating systems cannot be used in freezing temperatures, limiting the area in which solar technology can be used. Some types of solar water heaters cannot work as well or at all when it is cloudy. Because of the variations in temperatures and sunlight in most parts of the world, solar water heating systems sometimes need a backup water heating system to ensure the availability of hot water at all times. For many consumers, businesses, and institutions, this situation often means the purchase or use of a whole other hot water heater or water storage system with a means of keeping the water warm.

Impact of solar water heating systems

The use of a solar hot water heating system is positive for the environment. Using these systems reduces the amount of oil-based electricity used, resulting in fewer pollutants and lower greenhouse gas emissions. The manufacture of the elements in a solar water heating system can potentially affect the environment negatively, since most manufacturing processes require fossil fuels.

On an economic level, the installation and use of a solar water heating system can immediately save a consumer, business, or institution money in electricity costs. It only takes a few years for the system to pay for itself through energy cost savings. For example, a swimming pool solar water heater can pay for itself in about three years. However, some solar water heating systems, primarily those that heat swimming pools, are usually not eligible for any type of tax credit, rebate, or incentive for use.

Issues, challenges, and obstacles of solar water heating systems

Solar water heating systems can save money and are widely available. There are even "do it yourself" kits that allow the average home owner to add a solar hot water heater to his or her home. These kits are usually for batch solar water heaters. Despite this wide availability, these systems are not yet commonly used. In general, solar water heating systems can be expensive when compared to conventional water heating systems.

Advances in other water heating technologies also have drawn consumers away. There are new technologies that use natural gas to both heat water and spaces inside a home very efficiently. Such developments can potentially lengthen the payback time of a solar water heating system, making them less attractive to consumers and businesses.

PHOTOVOLTAIC CELLS

Photovoltaic cells, also known as solar cells, photoelectric cells, or just PV cells, are a type of solar technology that takes the energy found in light and directly converts it to electrical energy. PV cells are modular. That is, one can be used to make a very small amount of electricity, or many can be used together to make a large amount of electricity. A 3.9-inch (10-centimeter) diameter PV cell can make about one watt of power if the sun is directly overhead and the conditions are clear.

Because each photovoltaic cell produces only about one-half volt of electricity, cells are often mounted together in groups called modules. Each module holds about forty photovoltaic cells. By being put into modules, the current from a number of cells can be combined. PV cells can be strung together in a series of modules or strung together in a parallel placement to increase the electrical output.

When ten PV cell modules are put together, they can form an arrangement called an array or array field. Like modules, arrays can also be organized in a series or placed in parallel fashion. Arrays can be used to make electricity for a building or home. If many arrays are combined, they can create enough power to power a power plant. Some arrays are combined with a sun tracking device to ensure the sun hits the PV cell arrays throughout the day.

Even with photovoltaic cells, concentrating systems can be used to get more sunlight on the actual cells and help them produce more power. Such systems use mirrors or lenses to focus more sunlight on the PV cells. They also must be able to track the sun and be able to remove excess heat. If the temperature is too high in the PV cells, the amount of power each cell puts out is decreased.

Inside a photovoltaic cell are thin layers of a semiconductor material. Most commonly, these materials are silicon (melted sand) or cadmium telluride. The layers have a tiny amount of doping agent. Doping agents are impurities intentionally introduced in a chemical manner. Germanium and boron are examples of one type of doping agent that is used. The doping agents are important because they give the semiconductor materials the ability to make an electric current when exposed to light. These layers are stacked together. Each PV cell converts about 5 to 15 percent of the sunlight that hits it into electrical current.

Types of photovoltaic cells

There are several types of PV cells. A monocrystalline PV cell is blue or gray-black in color. At the rounded corner of each cell is a white backing. This backing shows through and makes a pattern that is easy to see. Some people do not use monocrystalline PV cells on their home or businesses because of their appearance. A module of PV cells is usually covered with tempered glass and surrounded by an aluminum frame.

A polycrystalline PV cell looks a little different than a monocrystalline PV cell. Polycrystalline PV cells are shaped like rectangles and colored sparkling blue. There is no white background showing. Thus, these PV cells look more uniform in appearance. Like monocrystalline cells, they are often covered in tempered glass and placed in an aluminum frame.

Another type is the amorphous or thin-film cell. However, this type of PV cell is less durable, not as efficient for the conversion of sunlight into power, and not as commonly used at this time. However, many experts believe that thin-film cells are the future of PV cell technology because they use less semiconductor material, do not need as much energy to manufacture, and are easier to mass produce than other PV cells.

Sometimes, photovoltaic systems have other components to make them useful for providing electricity. Two such components are an inverter and a storage device. The inverter helps change the DC power (direct current) produced by the cells to the AC (alternating current) used by most equipment, homes, and businesses that run on electricity in the United States.

The storage unit stores the energy created by the photovoltaic cells for use when there is little or no sun. One storage unit that

Solar Races and Other Contests

To encourage the development of solar energy and related technologies in the United States and around the world, there are a number of solar energy contests and competitions for students of many ages. Arguably the best-known solar energy competitions are the long-running solar car races. There is a World Solar Challenge, as well as smaller competitions such as the North American Solar Challenge. Sometimes cars compete in both races.

These solar cars are designed, built, and raced by college students who represent their school in the race. Students are trying to build the car that most effectively converts sunlight into energy and can travel the fastest on the route, but also last the longest in the race. They use solar collectors or PV cells to power the cars. Mechanical failures are common and have to be fixed on site. Students must also attract corporate sponsors to help pay for the cars and the travel involved in getting to and from the races.

These solar races have been held for a number of years. The first World Solar Challenge was held in 1987 in Australia. The North American Solar Challenge began in 2001. In 2005 twenty-eight teams competed in the North American Solar Challenge. That race ran from Austin, Texas, to Calgary, Alberta, Canada. The route was over 2,500 miles (4,100 kilometers) and took two weeks to complete. The route differs year to year. The University of Michigan won the 2005 North American Solar Challenge with a time of 53 hours, 59 minutes, and 43 seconds. The car averaged a speed of 46.2 miles (74.3 kilometers) per hour. Each year the speed the cars in the contest can achieve increases.

There are other solar contests. In 2005 the second annual Solar Decathlon was held on the National Mall and other locations in Washington, D.C. This contest is sponsored by the U.S. Department of Energy, the National Renewable Energy Laboratory, and private sponsors such as Home Depot. Groups of college students compete in events such as building the best solar house.

works well with photovoltaic cells is a battery, which stores the energy created electrochemically. The energy created by PV cells can also be stored as potential energy. Pumped water and compressed air are two types of potential energy. All of these storage types are used where the PV cells are located.

Current and future uses of photovoltaic cells

The first use for the first practical PV cell was a source of electricity for satellites orbiting Earth. PV cells were chosen because they were considered safer than nuclear power, another option being considered. On Earth, photovoltaic cells are used to

make electricity in places not connected to the power grid or where it is too costly to use electricity produced by the grid. This often happens in remote areas.

People who live in isolated houses or who want to be independent of the power grid use PV cells to provide electricity for their homes because of their adaptability. PV cells can power most household appliances, such as televisions, refrigerators, and computers, and they can also power electric fences and feeders for livestock. Photovoltaic systems can be used on farms to power pumps that provide water for livestock on grazing areas that are far away from the main farm.

Other independent, often isolated, objects use PV cells in similar ways. Navigation beacons can be powered by PV cells, as can remote monitoring equipment stations for pipeline systems, water quality systems, and meteorological information. Many traffic signals, street signs, billboards, bus stop lights, highway signs, security lighting, and roadside emergency telephones also use this technology.

Photovoltaic cells and modules are being integrated into buildings and homes to provide power. They usually supplement other

The solar car from The University of Calgary, Canada, leaves the starting point in Darwin on Sunday, September 25, 2005, in the 8th World Solar Challenge. Twenty-two solar-powered cars from the United States, France, Japan, and Canada will be trying to beat the Dutch team, which has won the last two events, in the 1,877 mile (3,021 kilometer) journey to Adelaide. © *David Hancock/Handout/EPA/Corbis.*

A man tends to one of the world's largest solar power plants. Each panel measures 80 by 160 centmeters and is part of 33,500 modules that form a solar power station providing five megawatts of electricity for about 2,000 households. ©*Waltraud Grubitzsch/EPA/Corbis.*

forms of power. There are incentives that will increase over time in many parts of the United States to use this technology. PV cells can also be used on the electrical grid in a supporting role for the transmission and distribution of power.

There are large photovoltaic systems that allow certain companies to avoid the electrical grid entirely. There have also been experiments to make large central power plants based on PV cells. However, PV cells have not yet proven to be cost effective in these situations. They are not yet efficient enough to justify the high cost of putting the project together and getting it started. If PV cells continue to become less expensive, such projects might become more practical.

In the future, the idea of building integrated photovoltaics (BIPV) might catch on. In this system, PV cells would be integrated into building materials such as shingles on roofs, windows, skylights, and the covers of insulation materials to provide a source of electricity for the home or building constructed with them. PV cells might also provide auxiliary power to automobiles.

Though PV cells are being installed for home or business use, they are not expected to be used on a widespread basis until 2010 at the earliest. By that time the cost of the technology is expected to be similar to the cost of electricity from the grid.

Benefits and drawbacks of photovoltaic cells

The use of photovoltaic cells has many positive aspects. They make no noise, require little to no maintenance, and are reliable. No special training is needed to operate a PV cell system. In addition, PV cells can be made a variety of sizes from very small to very large, providing flexibility in use. Moreover, many PV cells can be used anywhere because they can use both direct sunlight and diffuse sunlight. Finally, PV cell systems are long-lasting, maintaining their effectiveness for twenty to thirty years. Thus, they produce much more energy through their operation over their lifetime than is used to manufacture them.

Like many solar energy technologies, however, one major drawback to photovoltaic cells is that no power is produced when there is no sunshine. If the weather is poor and the sun is blocked, as when it rains or snows, these cells do not produce power. Photovoltaic cells also do not produce power at night. Because of this situation, some sort of backup system or alternate power supply is needed.

While the PV cells are very efficient producers of power, the manufacture of these cells does come at a significant energy cost. Also, over time the PV cells slowly become less efficient. At some point the cells lose most of their ability to be conductive. The costs of PV cells have remained high, though the prices have gone down over time. But because of the cost, the electricity PV cells produces costs more than electricity from the power grid in most areas.

Environmental impact of photovoltaic cells

While the widespread use of PV cells will reduce global warming by helping to cut down on the use of fossil fuel-created electricity, the manufacture of this solar technology can be polluting. Most manufacturers use mercury to construct solar cells. This is toxic waste that must be disposed of during their manufacture and after PV cells have reached the end of their usefulness.

Economic impact of photovoltaic cells

On a house-by-house level, photovoltaic cell systems are currently only cost-effective if the home is far away from power lines or if it is too costly to bring power lines to the house. The technology is still too expensive to be used everywhere on this level.

Solar Electric Light Fund

Founded in 1990, the Solar Electric Light Fund (SELF) strives to "promote, develop, and facilitate solar rural electrification and energy SELF-sufficiency in developing countries." By 2005 the fund had completed six separate projects on four continents and was working on several others. One such project in northern Nigeria used solar power to generate electricity for essential services such as water pumps to supply rural villages with fresh drinking water, lights for medical clinics and schools, and streetlights. All of the SELF projects used PV cell technology.

Though PV cells are costly, many governments and companies believe in the technology. Much money has been spent on research from the early 1990s to early 2000s. For instance, although BP Amoco is an oil company, it has invested in producing PV technology. By 2000 its goal was to become the biggest producer of PV cells in the world. Amoco has already put some PV cells in its gas stations.

Because of this economic support of PV cell research, an industry has grown up around them. In 2004 the production worldwide of photovoltaic cells increased by 60 percent, and this growth is expected to continue. Manufacturing costs have declined every year for several years. By 2010 it is expected that the PV market may be $30 billion worldwide, perhaps making it one of the big growth industries in the world. As the market expands and research into better technology grows, prices will likely come down. Thus, the future of PV cells is extremely promising.

Societal impact of photovoltaic cells

The use of PV cells can increase the availability of electricity around the world. Photovoltaic cells have brought power to parts of the world that did not have power before, except from generators powered by diesel fuel. Developing countries can best benefit from PV cell technology. The World Bank has installed PV systems in developing countries to provide a source of electricity. By 2001 at least 500,000 of the systems have been put in countries such as Sri Lanka, Indonesia, Kenya, Mexico, and China. China has 100,000 of the systems, while Kenya has 150,000. These numbers are expected to increase.

PV cells are also making an impact in developed countries such as the United States and Japan. By 1995 photovoltaic cells and modules added a capacity supply of 4.6 megawatts to the U.S. power grid. As of 2001 at least 200,000 residences in the United States used PV technology in some form.

Issues, challenges, and obstacles of photovoltaic cells

The use of photovoltaic cells can be challenging. Since the electricity they produce is DC and most applications of electric power use AC, a power conditioning system is needed to ensure that the DC is converted to AC and is safe to use.

Another factor that has limited the widespread use of PV cells, especially to make large amounts of electricity, is the PV cell system's efficiency. PV cells are not particularly efficient in the amount of sunlight that is converted to electricity. If PV cells can turn more than 15 percent of the sunlight's energy into electricity, they will become an even more attractive alternative to electricity created by fossil fuels.

DISH SYSTEMS

The dish system is also known as the distributed-point-focus system. Dish systems feature small, parabolic mirrors that are dish-shaped. They reflect the sunshine onto a receiver. A two-axis tracking system is employed to move the mirrors to ensure that as much solar energy reflected by the mirrors is captured as possible. The receiver is usually mounted above the mirrors at the center of the dish, its focal point. Inside the receiver is a fluid, which transfers the intense heat created by focusing the sunlight on the receiver. This makes electricity. Each dish can produce from 5 to 50 kilowatts of electricity. The dishes can be used singly or linked together.

Dish systems can be part of another solar technology called a dish-engine system. The dish part of the system is similar to the one described above. But the dish-engine system also includes an engine. The receiver in this system transfers the sunlight's energy to the engine. The engine, often one that can be driven by an external heat source, converts the energy to heat. The heat is then made into mechanical power. This happens by the compression of the working fluid, like steam, with the heat. It is then expanded via a turbine or piston. After mechanical power is produced, an electric generator or alternator turns the mechanical power into electrical power.

A dish-engine system can also be linked. If they are linked, they can potentially produce a significant amount of electricity. Ten

Tracking parabolic solar dishes concentrate incoming solar radiation to a central point, where a thermal collector captures the heat and transforms it into energy. © *Otto Rogge/Corbis.*

25-kilowatt dish-engine systems can produce 250 kilowatts of power. This would only require an acre of land.

Current uses of dish systems

Dish systems and dish-engine systems are used to generate electric power. However, they are still in the experimental and demonstration phases. The most electricity that has been produced from a single dish-engine system is about 50 kilowatts. More commonly, as of 2005 each system generates about 25 kilowatts.

It is believed that linked dish-engine systems will be a significant electricity producer of the future. Dish-engine systems can also be hybrids. That is, they might be combined with natural gas into a hybrid that can ensure the constant production of electricity.

Because of the size of the dishes involved, they must be used on a significant scale. They are not made for just one home. In 2004 a dish made by Stirling Energy that could produce 25 kilowatts of electricity was 38 feet (11.5 meters) across and 40 feet (12 meters) tall. It is expected that such systems will be produced on a commercial scale. The Arizona Public Service Company has already agreed to buy ten such systems to make power. Other southwestern states in the United States are also considering purchasing them.

Benefits and drawbacks of dish systems

Dish systems and dish-engine systems are efficient producers of electricity. When they are linked together, they can produce more energy per acre than any other kind of solar energy technology. As the technology improves, they may be able to provide electricity for areas off the electricity grid or as an alternative to the electricity grid. Using technologies such as the dish system and dish-engine system can lead to less dependence on fossil fuels to make electricity.

To use the dish system and dish-engine system, however, very intense sunshine is needed. In the United States, the kind of sunshine needed can only be found in the southwestern part of the country. Key to the use of dish-engine systems is space. If such systems are going to be used on any type of scale, large amounts of empty space are needed for the many dishes to operate.

Dish systems and dish-engine systems also need more maintenance than other types of solar energy technologies. There are many moving parts, especially if a generator or motor is attached, which could break down and disrupt the flow of electricity.

Impact of dish systems

No matter if one or many dish systems and dish-engine systems are being used, the environment where they are placed will be affected. In the United States, the systems will most likely be placed in deserts, which means that previously barren deserts will be covered with technology. Wildlife and plant life in the area could be negatively affected. If dish-engine systems reach a commercial scale, this impact could be devastating. The very environmentalists who support solar energy might find themselves at odds with the reality of the technology.

Economically, if this technology reaches maturity, it will provide a potentially cheap alternative source of power. This could affect how electric companies and energy providers run their businesses. It could also result in lower energy costs for consumers.

On a societal level, dish systems and dish-engine systems could provide a source of electricity for developing countries located in extremely sunny environments. The availability of such electricity could improve quality of life there.

Issues, challenges, and obstacles of dish systems

It is unclear if the environmental issues related to the use of the dish system and dish-engine system will negatively affect the use of these systems as a widespread source of electricity. A balance must be created between the effect on the environment and the creation of electricity by such alternative forms of energy.

TROUGH SYSTEMS

The trough system, also called the line-focus collector, focuses sunlight to create electricity. The trough system has its name because each collector is shaped like a trough that is parabolic (curved) in shape. There is a tube running down the middle of the trough with fluid inside. Mirrors inside the trough concentrate sunlight on that tube and heat the fluid inside it. The fluid is usually dark oil, but other substances can be used. The oil can get as hot as 752°F (400°C). The heat from the oil is transferred to water, which turns into steam. The steam can be used to power a turbine-generator or other machinery to produce the electricity.

Trough systems are modular. That means they can be linked together to make a larger amount of electricity than can be created by an individual trough. Many troughs together form a collector field when they are put in parallel rows. In a collector field the troughs are set in a certain way, usually aligned in an axis running from north to south. This allows the troughs to track the sun from east to west, the direction the sunlight moves during the day. An individual trough system can produce up to 80 megawatts of electricity.

There are several ways to make sure trough systems produce electricity after the sun goes down. Some trough systems have a means of thermal storage. That is, they can save the heat transfer fluid while still hot. By doing so, the troughs can still power the turbines after the sun goes down. However, trough systems are usually hybridized, meaning they are combined with a fossil fuel system for supplying electricity. Usually, the heat is created by natural gas. Using a gas-powered steam boiler is also possible. If trough systems are hybridized, they can produce power at all times. Coal-powered plants can also be supplemented by the trough system.

Current uses of trough systems

The trough system is already being used to make electricity around the world. As of 2001 these types of systems accounted for 90 percent of the solar energy-produced electricity in the

Parabolic trough mirrors at a solar power plant. © *Royalty-Free/Corbis.*

world. Since the early 1990s troughs have been operating in Southern California's Mojave Desert. These troughs have provided as much as 354 megawatts of electricity for the power grid in the Southern California area.

Benefits and drawbacks of trough systems

Trough systems have many benefits, which is why they have been so widely adopted. Except for the generator, trough systems require minimal maintenance. They are also very flexible in terms of how many or few troughs can be linked together. The energy they produce is not quite on the price level of fossil fuel-produced electricity, but the figure is often very close.

As with all solar energy technologies, the fact that the sun does not shine at all times is a major drawback. For trough systems to operate to capacity, they need intense, direct sunshine. Such sunshine can only be found in the United States in the desert Southwest. Trough systems also take up a significant amount of space when they are linked together to provide power on a widespread scale.

Impact of trough systems

While the trough system produces pollutant-free energy, many systems used together can take up much land. They are often placed in a desert that had previously been free of buildings or other structures. Placing a collector farm or any significant number of trough systems may litter this landscape and potentially destroy it. Animals and plants in the area could be negatively affected by the presence of this technology.

Despite the environmental costs, many governments support the use of trough systems to generate power. There are federal tax incentives for the use of trough systems. The State of California, for example, has mandated that power made by renewable energy sources must be purchased, and this is one technology the state has encouraged. If trough systems are ever used on a widespread basis, they could provide a cheap alternative source of power.

Issues, challenges, and obstacles of trough systems

Because the trough system is a more commonly used technology than other types of solar energy, there is a familiarity with it in the energy industry. This awareness makes it more appealing. If trough systems can spread to all sunny parts of the world, solar energy in general technology could become more accepted. However, the space requirement of the trough system will limit the growth of this industry.

SOLAR PONDS

A solar pond is a large, controlled body of water that collects and stores solar energy. Solar ponds do not use tracking systems such as mirrors, nor do they concentrate the sun's rays like many other solar energy technologies.

There are two types of convecting solar ponds. (Convection is a process in which a fluid such as water circulates, and in so doing the circulation causes a transfer of heat.) One is called a salt-gradient pond. At the very bottom of the pond is a dark layer that can absorb heat. This is usually a liner made of butyl rubber or other dark material. In addition to helping the water absorb the heat, it helps protect the nearby soil and groundwater from being contaminated by the saltwater from the solar pond.

In the pond, there is a significant amount of salt located near the bottom. The types of salt commonly used are sodium chloride or magnesium chloride. The water is saturated (filled entirely) or almost saturated with salt. The closer to the surface, the less salt is found in the water. At the very top of the pond is a layer of

Salt evaporation ponds at Shark Bay, Western Australia. © *Sergio Pitamitz/ Corbis.*

freshwater (that is, water without salt). This change in saltiness forms layers in the pond. The gradual change in the amount of salt is called a salt-density gradient.

The layers of saltwater stop the natural tendency of hot water to rise to the surface. Thus, the water that is heated by the sun stays at the bottom of a solar pond. The layers that are close to the surface remain cool. There is a significant temperature difference between the top and the bottom of a solar pond, though some heat can be stored on every layer. Temperatures as high as 179 to 199°F (82 to 93°C) can be found at the bottom.

The heat is extracted by a heat exchanger at the bottom of the pond. This heat energy can power an engine, provide space heating, or produce electricity via a low-pressure steam turbine. The heated saltwater can be pumped to the location where the heat is needed. After the heat is used, the water can be returned to the solar pond and heated again.

The second type of convecting pond is a membrane pond. A membrane pond is similar to the salt-gradient pond except the

Ocean Thermal Energy Conversion

The concept behind solar ponds can be applied in the ocean in ocean thermal energy conversion (OTEC). In the ocean the water has different temperatures at different depths. It is often warm on the surface and colder the farther from the surface it is. If the temperature difference is at least 68°F (20°C), such as in tropical areas where the ocean is deep, then OTEC could be used to create energy.

To take advantage of OTEC, a pipe would be used to pump a significant amount of water to the surface. There, it would be run through a heat exchanger to capture the energy. In addition to providing electricity, the system could be adapted to produce freshwater. It could also be used to provide water full of nutrients in which such food items as fish and vegetables could be raised.

A prototype of OTEC was used in Hawaii in the mid-1990s. In the future, developing countries in coastal tropical areas could employ the technology.

layers of water are physically divided. They are separated by membranes that are thin and transparent. The separation of layers physically prevents convection (circulating movement). With a membrane pond the heat that is created is also removed from the bottom layer of the pond as in a salt-gradient pond.

There are also two types of nonconvecting ponds. One is called a shallow solar pond. This pond has no saltwater. Pure freshwater is kept inside a large bag. The bag allows convection to take place but limits the amount of water that can be evaporated. At the bottom of the bag is a black area. Foam insulation can also be found near the bottom. On top of the bag are two types of glazing. These glazings are usually sheets of plastic or glass.

In a shallow solar pond, the sunshine heats the bag and the water inside during the day. The heat energy is extracted at night. The heated water is pumped into a large heat storage tank. This process can be difficult because heat loss is possible. The problems with heat loss have meant that shallow solar ponds have not been fully developed as a technology.

The other type of nonconvecting pond is the deep, saltless pond. The primary difference between this pond and the shallow solar pond is that the water is not pumped in and out of its storage medium. This limits the amount of heat that can be lost.

Current and future uses of solar ponds

Solar ponds can be used in a number of ways. They can make electricity or be used to provide heating for community, residential, and commercial purposes. They can also provide low-temperature heat for certain industrial and agricultural purposes, and they can also be used in preheating applications for industrial processes that require higher temperatures. In addition, solar ponds can be used to desalinate (remove the salt from) water. In Australia a pond at the Pyramid Hill salt works in Northern Victoria is used by the company to help make salt.

Solar ponds have been used for several decades. In the 1970s in Israel, a salt-gradient pond was created near the Dead Sea. Until 1989 it generated 5 megawatts of electricity. The project ended because of the high costs involved. Similar systems were built in California and other locations in the United States as well as India and Australia, though they were on a smaller scale. Several shallow solar ponds were built by the Tennessee Valley Authority.

There are a number of potential applications for solar ponds. Such ponds might be used to grow and farm brine shrimp or other sea creatures that are used as feed for livestock. In Australia solar pond projects are planned that would dry fruit and grain. Some researchers hope to use solar ponds in the production of dairy products.

Benefits and drawbacks of solar ponds

Solar ponds are very versatile. They can use both direct sunlight as well as diffuse radiation on cloudy days. They can store the heat they collect during the daytime hours for use at night. A separate thermal storage unit is not always needed.

Another benefit is that solar ponds can be used in nearly any climate. They can even be used in winter when the top layer of a salt-gradient pond becomes covered in ice. They are also reusable: The water from which the heat is removed can be returned to the pond.

Finally, solar ponds do not always cost much to construct. There is no solar collector that needs to be cleaned. Because the solar pond can be built to be big, large amounts of power can be produced.

One drawback is that solar ponds require a very large area of flat land. It can be difficult to find the empty land needed to make the pond big enough to be used. In addition, lots of salt is also needed.

Make Your Own Solar Pond

A small solar pond is easy to make at home with an aquarium, some food coloring, a lamp with a 100-watt light bulb, some salted water, and a few other items. First, set up a small five-gallon aquarium. Take two gallons of warm water and mix in one cup of salt. Mix until all the salt is dissolved. Then add in another one-third cup of salt. Let the mixture cool. Mix in a little red food coloring and put the mixture in the aquarium.

Take a very small funnel and a foot-long piece of hose. Attach the hose to the bottom of the funnel. Put the hose about half way into the water in the tank. Slowly add a gallon of freshwater. After the water is added, move the hose up toward the top of the tank without moving the hose above the water level. This step helps to create a gradient.

Now something small that can float is needed. Such items could be a plastic coffee can lid or a very thin wooden block. Put this item on the water. Pour water slowly onto the floating object. Leave the aquarium alone for one hour.

After one hour, put a few drops of blue food coloring onto the floating object. Then put the lamp over the tank and turn it on so the light is shining down. Put a thermometer that can go as high as 120 degrees Fahrenheit (48 degrees Celsius) in the water. Monitor the aquarium for the next twenty-four hours. The temperature will rise over that time period. As this solar pond heats up, three different colors will appear representing the three different levels of salty water that would be found in a solar pond.

Impact of solar ponds

Some of these ponds are very large, which can affect the environment around them. Measures must be taken to ensure that the salt from the solar ponds does not contaminate the soil. This contamination could very negatively affect the environment. Solar ponds can also have a positive environmental impact, however. When combined with desalting units, solar ponds can be used to purify water that is contaminated. Solar ponds make heat energy without burning any fuels and save conventional energy resources.

Despite the fact that solar ponds are not particularly efficient in their production of energy, they are inexpensive. However, they are not seen as economically advisable in the long term. As a result, there is very little commercial interest in them in most parts of the world.

Solar ponds can be a source of cheap salt in some countries. In Australia, for example, solar ponds can productively use lands that have too much salt in them to be used for anything else. All over Australia there are a number of underground sources of saltwater.

This water can be turned into freshwater using solar ponds in a profitable fashion. These uses could be positive for Australian society, resulting in the creation of new jobs, industries, and sources of water.

Issues, challenges, and obstacles of solar ponds

While solar ponds have much potential, there has not been very much investment in the technology behind them. Yet the solar ponds could provide freshwater and electricity in coastal desert regions and islands. However, such applications have not yet been realized.

SOLAR TOWERS

Solar towers, also known as power towers, central receivers, or heliostat mirror power plants, use solar energy to generate enough power to provide electricity over a large area. In this system the sun's power is collected by a large field of flat, movable mirrors. Sometimes there are thousands of mirrors. The mirrors, called heliostats, move so they can track the sun. They are focused on one single, fixed receiver that is located on top of a tall, central tower. Temperatures can be produced from 1,022 to 2,732°F (550 to 1,500°C) at the receiver.

The receiver collects all the energy and heat into a heat-transfer fluid that is flowing through it. In early power towers, this fluid was plain water. However, more recent models usually use molten salt, though liquid sodium, nitrate salt, and oil are also used. The heat energy held in the salt is used to boil water and make steam. This steam is used to generate electricity in a steam generator, usually located at the foot of the tower.

Molten salt can act as an efficient thermal storage medium for the heat collected in the solar tower. The heat can be stored for many hours or several days in this fashion. This storage medium is very important. It allows the solar towers to be operational for up to 65 percent of the year. The rest of the time, a backup fuel source is used. When there is no energy storage medium, solar towers can only be used for about 25 percent of the year.

Current and future uses of solar towers

In the 1970s supporters believed that solar tower technology would take off. A number of solar tower technologies were implemented in the successive decades. In California there have been several solar tower projects. Solar One, which operated from 1982 to 1988, used water as a heat-transfer fluid in the receiver. It used 1,818 mirrors placed in semicircles around a tower that was 255 feet (78 meters) high. The mirrors focused the sunlight onto a boiler at the top. The use of water created problems for storage of the heat created and for running the turbine. Solar One was remade in 1992 to replace the water with molten salt. Despite this change, Solar One only functioned for a short time longer.

California funded another solar tower project that required an initial investment of $150 million. Solar Two operated from 1996 to 1999, had 10 megawatts of capacity, and also used molten salt. The success of Solar Two showed that the technology could work on a commercial basis. Solar towers were built in other countries as well. In Spain a solar tower was built that was smaller than the power towers built in California. It was constructed in 1982 south of Madrid and could produce up to 50 kilowatts of power. It was only used on an experimental basis to heat air.

Despite this early promise, as of 2001 there were no commercial solar towers in operation anywhere in the world. But more projects are being planned. In the future it is believed that solar towers will be built that can provide power for from 100,000 to 200,000 homes. Future projects might include a project in Spain called Solar Tres ("Solar Three") that will also use molten salt. Solar Tres was not seen as a short-term experiment but a long-term source of power. South Africa is planning on building a solar tower plant as well.

The most ambitious solar tower project was planned in Australia. In the early 2000s the country talked about building a giant solar tower, one of the tallest structures in the world, out in the desert near Mildura, Victoria, Australia. It would be 0.62 miles (0.9 kilometers) high and would produce 650 gigawatts of electricity each year at its peak to serve 70,000 consumers or 200,000 homes. This tower would be connected to thirty-two turbines. The tower would cost at least US$720 million to build. Australia hopes to have the tower actually working in 2008, if funding and logistics can be worked out. It is unclear how it would be built. There were other issues such as how to protect it from high winds, if it would be commercially workable, and if it would be technologically out of date by the time it was completed.

Benefits and drawbacks of solar towers

Solar towers have one important advantage over other types of solar power: They continually generate electricity as long as they have a means of heat storage such as molten salt. This means that they can be used to provide reliable power for customers over a long period of time. However, there are many drawbacks to solar towers as well. The technology is currently very costly. It might cost too much to make the power when considering the cost of building the tower itself. Also, solar towers are not particularly efficient means of converting sunshine into electricity. Only about 1 percent of the sunlight that hits the tower is actually made into electricity. Moreover, the size of the tower makes it difficult to place.

Impact of solar towers

Solar towers take up a lot of space and are usually put in the desert or on empty land. The construction of such a large project could negatively affect the environment. The size and scope of what a solar tower looks like—the field of mirrors, the high tower, and the generator—could also negatively affect the location in which the tower is placed.

On the other hand, if this technology reaches maturity, solar towers could provide a cheap alternative source of power in the future. Although at present solar towers produce power that costs more than current electricity made with fossil fuels, as fossil fuels run out, the electricity made with fossil fuels will become more expensive and solar towers will become comparatively cheaper.

Issues, challenges, and obstacles of solar towers

Solar towers have many positive aspects. They can run for long periods of time on stored energy, which comes from the sun. This makes solar towers different from many other renewable energy technologies. Yet solar towers have not caught on as a power-producing technology. Perception of the potential of solar towers needs to change for it to be considered a viable electric-producing source in the future. As long as the technology continues to develop, solar towers have a chance to be an important source of renewable energy in the future.

SOLAR FURNACES

Like solar power towers, solar furnaces use mirrors to concentrate sunlight onto one point to achieve high temperatures. The solar energy is collected from over a wide area. Solar furnaces can create higher temperatures than solar towers. There are several types of solar furnaces, each of which produces a different wattage of power.

The best known solar furnace is called a high-flux solar furnace. It uses just one flat mirror or heliostat that is very large in size. It tracks the sun to ensure the greatest reflection of sunlight onto the primary concentrator. The concentrator consists of twenty-five or so individual curved mirrors. These mirrors focus the light, called a solar flux, at a target inside the building.

The large parabolic mirror, target area, and tower of the solar furnace in Odeillo, France. © *Paul Almasy/ Corbis.*

The light from the concentrator is focused on a circle or target inside the furnace. The focused beam of light created by the concentrator is much, much stronger than normal sunlight. At its focal point it can produce the energy of 2,500 suns. There can also be a reflective secondary concentrator added to the focus. The equivalent of up 20,000 suns can then be produced. When a refractive concentrator is added to the system to focus even more light on the beam, the intensity can equal an amazing 50,000 suns. Temperatures rise very rapidly in a solar furnace, more than 1,832°F (1,000°C) per second. The power level inside the furnace is adjustable by a device called an attenuator, which works like pulling down blinds over a window.

Current and future uses of solar furnaces

Solar furnaces are primarily used to generate heat or steam to make electricity and for industrial use. Steam created by solar furnaces can be used to run generators and industrial equipment. An advantage of using solar-created heat in such industrial processes is that the heat is clean, meaning that it produces no harmful emissions. The first solar furnace was designed in Germany in 1921. It used a parabolic concentrator and lenses. Several more were built in Germany, France, and the United States between the 1930s and 1950s. One built in France in 1952 could produce 50 kilowatt hours of electricity.

In 1970 one of the most powerful solar furnaces built was constructed. It was located in Odeillo, France, at one of the sunniest points in Europe. It can produce about 100 kilowatt hours of electricity and has the capability of making heat as hot as 59,432°F (33,000°C). On the hillside opposite the furnace are 9,600 to 11,000 flat mirrors over 1,860 square miles (4,817 square kilometers) that track the sun and reflect sunlight onto one side of the furnace. On this side of the ten-story furnace are curved mirrors that cover its face. These mirrors are joined together to act as one large mirror. They focus the sun's energy onto an area that is less than 10 square feet (1 square meter) in order to create the high temperatures. This solar furnace is primarily used for scientific experiments on high temperature applications.

As of 2005 there are only a few solar furnaces in working order. Besides the one in Odeillo, France, they include smaller ones in China and the United States. The solar furnace in the United States is located at the National Renewable Energy Laboratory's CSR (Concentrated Solar Radiation) User Facility in Golden, Colorado. A high-flux solar furnace, it was built in 1990 and puts out 10 kilowatts of power. It is used to experiment on how solar furnaces can be used in industry.

The future uses of solar furnaces are being determined by furnaces such as the one at the CSR User Facility. There, experiments are being conducted with ceramics, surface hardening, coatings, and processes related to the processing of silicon. It is believed that solar furnaces can be used in manufacturing in the production of aerospace products, defense products, and in electronics. Solar furnaces also could be used to break down and destroy toxic waste. In these uses the high-flux solar furnace would replace laser furnaces and furnaces using fossil fuels.

Solar furnaces also have the potential to be used in materials processing and materials manufacturing that require high temperatures. The furnaces can quicken the pace of weathering for studying of future materials and how they will change over time. The CSR furnace can weatherize an object the equivalent of twenty years in only two-and-one-half months.

Benefits and drawbacks of solar furnaces

A solar furnace can produce very high temperatures for industrial processes without the environmental costs and economic costs related to fossil fuels. Because some scientific experiments need a more pure fuel than is possible with fossil fuels, solar furnaces could be used because of the purity of sunlight. One drawback of solar furnaces is that they are very large and costly to build. They require a large amount of land in a sunny area to be effective.

Impact of solar furnaces

Constructing a solar furnace has a profound effect on the environment in which it is placed. Acres, if not many square miles, of land are needed to place mirrors and a furnace, as well as any related industrial equipment. The building and operation of a solar furnace affects the local wildlife and local plant life.

On an economic level, it is unclear if the cost of building a solar furnace is cost-effective based on how much energy is produced by such furnaces. However, if solar furnaces prove to be cost-effective and feasible in more than a few areas, they could provide an alternative heat and electricity source for many types of industry. The growth in the use of solar furnaces could be an ideal way for industry to convert to solar energy and use less fossil fuels.

Issues, challenges, and obstacles of solar furnaces

Solar furnace technology has existed for many years but never has been fully explored or used on a widespread commercial basis. It is unclear if solar furnaces will ever be used on any type of scale because of the limitations in their placement and use. However, the research happening at the CSR User Facility and others like it could lead to breakthroughs that improve the technology and/or lower the cost.

For More Information

Books

Buckley, Shawn. *Sun Up to Sun Down: Understanding Solar Energy.* New York: McGraw-Hill, 1979.

McDaniels, David K. *The Sun.* 2nd ed. New York: John Wiley & Sons, 1984.

Periodicals

Brown, Kathryn. ''Invisible Energy.'' *Discover* (October 1999): 36.

Corcoran, Elizabeth. ''Bright Ideas.'' *Forbes* (November 24, 2003): 222.

Dixon, Chris. ''Shortages Stifle a Boom Time for the Solar Industry.'' *New York Times* (August 5, 2005): A1.

Libby, Brian. ''Beyond the Bulbs: In Praise of Natural Light.'' *New York Times* (June 17, 2003): F5.

Nowak, Rachel. ''Power Tower.'' *New Scientist* (July 31, 2004): 42.

Pearce, Fred. ''Power of the Midday Sun.'' *New Scientist* (April 10, 2004): 26.

Perlin, John. ''Soaring with the Sun.'' *World and I* (August 1999): 166.

Provey, Joe. "The Sun Also Rises." *Popular Mechanics* (September 2002): 92.

Tompkins, Joshua. ''Dishing Out Real Power.'' *Popular Science* (February 1, 2005): 31.

Web sites

The American Solar Energy Society. http://www.ases.org/ (accessed on September 1, 2005).

''Clean Energy Basics: About Solar Energy.'' National Renewable Energy Laboratory. http://www.nrel.gov/clean_energy/solar.html (accessed on August 25, 2005).

''Conserval Engineering, Inc.'' American Institute of Architects. http://www.solarwall.com/ (accessed on September 1, 2005).

''Florida Solar Energy Center.'' University of Central Florida. http://www.fsec.ucf.edu (accessed on September 1, 2005).

''Photos of El Paso Solar Pond.'' University of Texas at El Paso. http://www.solarpond.utep.edu/page1.htm (accessed on August 25, 2005).

''Solar Energy for Your Home.'' Solar Energy Society of Canada Inc. http://www.solarenergysociety.ca/2003/home.asp (accessed on August 25, 2005).

The Solar Guide. http://www.thesolarguide.com (accessed on September 1, 2005).

''Solar Ponds for Trapping Solar Energy.'' United National Environmental Programme. http://edugreen.teri.res.in/explore/renew/pond.htm (accessed on August 25, 2005).

Where to Learn More

BOOKS

Angelo, Joseph A. *Nuclear Technology.* Westport, CT: Greenwood Press, 2004.

Avery, William H., and Chih Wu. *Renewable Energy from the Ocean.* New York: Oxford University Press, 1994.

Berinstein, Paula. *Alternative Energy: Facts, Statistics, and Issues.* Phoenix, AZ: Oryx Press, 2001.

Boyle, Godfrey. *Renewable Energy,* 2nd ed. New York: Oxford University Press, 2004.

Buckley, Shawn. *Sun Up to Sun Down: Understanding Solar Energy.* New York: McGraw-Hill, 1979.

Burton, Tony, David Sharpe, Nick Jenkins, and Ervin Bossanyi. *Wind Energy Handbook.* New York: Wiley, 2001.

Carter, Dan M., and Jon Halle. *How to Make Biodiesel.* Winslow, Bucks, UK: Low-Impact Living Initiative (Lili), 2005.

Cataldi, Raffaele, ed. *Stories from a Heated Earth: Our Geothermal Heritage.* Davis, CA: Geothermal Resources Council, 1999.

Close, Frank E. *Too Hot to Handle: The Race for Cold Fusion.* Princeton, NJ: Princeton University Press, 1991.

Cook, Nick.*The Hunt for Zero Point: Inside the Classified World of Antigravity Technology.* New York: Broadway Books, 2003.

Cuff, David J., and William J. Young. *The United States Energy Atlas,* 2nd ed. New York: Macmillan, 1986.

Dickson, Mary H., and Mario Fanelli, eds. *Geothermal Energy: Utilization and Technology.* London: Earthscan Publications, 2005.

Domenici, Peter V. *A Brighter Tomorrow : Fulfilling the Promise of Nuclear Energy.* Lanham, MD: Rowman and Littlefield, 2004.

Ewing, Rex. *Hydrogen: Hot Cool Science—Journey to a World of the Hydrogen Energy and Fuel Cells at the Wassterstoff Farm.* Masonville, CO: Pixyjack Press, 2004.

Freese, Barbara. *Coal: A Human History*. New York: Perseus, 2003.

Frej, Anne B. *Green Office Buildings: A Practical Guide to Development.* Washington, DC: Urban Land Institute, 2005.

Gelbspan, Ross. *Boiling Point: How Politicians, Big Oil and Coal, Journalists and Activists Are Fueling the Climate Crisis*. New York: Basic Books, 2004.

Geothermal Development in the Pacific Rim. Davis, CA: Geothermal Resources Council, 1996.

Graham, Ian. *Geothermal and Bio-Energy*. Fort Bragg, CA: Raintree, 1999.

Heaberlin, Scott W. *A Case for Nuclear-Generated Electricity: (Or Why I Think Nuclear Power Is Cool and Why It Is Important That You Think So Too)*. Columbus, OH: Battelle Press, 2003.

Howes, Ruth, and Anthony Fainberg. *The Energy Sourcebook: A Guide to Technology, Resources and Policy*. College Park, MD: American Institute of Physics, 1991.

Husain, Iqbal. *Electric and Hybrid Vehicles: Design Fundamentals.* Boca Raton, FL: CRC Press, 2003.

Hyde, Richard. *Climate Responsive Design.* London: Taylor and Francis, 2000.

Kaku, Michio, and Jennifer Trainer, eds. *Nuclear Power: Both Sides.* New York: Norton, 1983.

Kibert, Charles J. *Sustainable Construction: Green Building Design and Delivery*. New York: Wiley, 2005.

Leffler, William L. *Petroleum Refining in Nontechnical Language.* Tulsa, OK: Pennwell Books, 2000.

Lusted, Marcia, and Greg Lusted. *A Nuclear Power Plant.* San Diego, CA: Lucent Books, 2004.

Manwell, J. F., J. G. McGowan, and A. L. Rogers. *Wind Energy Explained.* New York: Wiley, 2002.

McDaniels, David K. *The Sun.* 2nd ed. New York: John Wiley & Sons, 1984.

Morris, Robert C. *The Environmental Case for Nuclear Power.* St. Paul, MN: Paragon House, 2000.

National Renewable Energy Laboratory, U.S. Department of Energy. *Wind Energy Information Guide.* Honolulu, HI: University Press of the Pacific, 2005.

Ord-Hume, Arthur W. J. G. *Perpetual Motion: The History of an Obsession.* New York: St. Martin's Press, 1980.

Pahl, Greg. *Biodiesel: Growing a New Energy Economy.* Brattleboro, VT: Chelsea Green Publishing Company, 2005.

Rifkin, Jeremy. *The Hydrogen Economy.* New York: Tarcher/Putnam, 2002.

Romm, Joseph J. *The Hype of Hydrogen: Fact and Fiction in the Race to Save the Climate.* Washington, DC: Island Press, 2004.

Seaborg, Glenn T. *Peaceful Uses of Nuclear Energy.* Honolulu, HI: University Press of the Pacific, 2005.

Tickell, Joshua. *From the Fryer to the Fuel Tank: The Complete Guide to Using Vegetable Oil as an Alternative Fuel.* Covington, LA: Tickell Energy Consultants, 2000.

Wohltez, Kenneth, and Grant Keiken. *Volcanology and Geothermal Energy.* Berkeley: University of California Press, 1992.

Wulfinghoff, Donald R. *Energy Efficiency Manual: For Everyone Who Uses Energy, Pays for Utilities, Designs and Builds, Is Interested in Energy Conservation and the Environment.* Wheaton, MD: Energy Institute Press, 2000.

PERIODICALS

Anderson, Heidi. "Environmental Drawbacks of Renewable Energy: Are They Real or Exaggerated?" *Environmental Science and Engineering* (January 2001).

Behar, Michael. "Warning: The Hydrogen Economy May Be More Distant Than It Appears." *Popular Mechanics* (January 1, 2005): 64.

Brown, Kathryn. "Invisible Energy." *Discover* (October 1999): 36.

Burns, Lawrence C., J. Byron McCormick, and Christopher E. Borroni-Bird. "Vehicles of Change." *Scientific American* (October 2002): 64-73.

Corcoran, Elizabeth. "Bright Ideas." *Forbes* (November 24, 2003): 222.

Dixon, Chris. "Shortages Stifle a Boom Time for the Solar Industry." *New York Times* (August 5, 2005): A1.

Feldman, William. "Lighting the Way: To Increased Energy .Efficiency." *Journal of Property Management* (May 1, 2001): 70.

Freeman, Kris. "Tidal Turbines: Wave of the Future?" *Environmental Health Sciences* (January 1, 2004): 26.

Graber, Cynthia. "Building the Hydrogen Boom." *OnEarth* (Spring 2005): 6.

Grant, Paul. "Hydrogen Lifts Off—with a Heavy Load." *Nature* (July 10, 2003): 129-130.

Guteral, Fred, and Andrew Romano. "Power People." *Newsweek* (September 20, 2004): 32.

Hakim, Danny. "George Jetson, Meet the Sequel." *New York Times* (January 9, 2005): section 3, p. 1.

Lemley, Brad. "Lovin' Hydrogen." *Discover* (November 2001): 53-57, 86.

Libby, Brian. "Beyond the Bulbs: In Praise of Natural Light." *New York Times* (June 17, 2003): F5.

Linde, Paul. "Windmills: From Jiddah to Yorkshire." *Saudi Aramco World* (January/February 1980). This article can also be found online at http://www.saudiaramcoworld.com/issue/198001/windmills-from.jiddah.to.yorkshire.htm.

Lizza, Ryan. "The Nation: The Hydrogen Economy; A Green Car That the Energy Industry Loves." *New York Times* (February 2, 2003): section 4, p. 3.

McAlister, Roy. "Tapping Energy from Solar Hydrogen." *World and I* (February 1999): 164.

Motavalli, Jim. "Watt's the Story? Energy-Efficient Lighting Comes of Age." *E* (September 1, 2003): 54.

Muller, Joann, and Jonathan Fahey. "Hydrogen Man." *Forbes* (December 27, 2004): 46.

Nowak, Rachel. "Power Tower." *New Scientist* (July 31, 2004): 42.

Parfit, Michael. "Future Power: Where Will the World Get Its Next Energy Fix?" *National Geographic* (August 2005): 2–31.

Pearce, Fred. "Power of the Midday Sun." *New Scientist* (April 10, 2004): 26.

Perlin, John. "Soaring with the Sun." *World and I* (August 1999): 166.

Port, Otis. "Hydrogen Cars Are Almost Here, but . . . There Are Still Serious Problems to Solve, Such As: Where Will Drivers Fuel Up?" *Business Week* (January 24, 2005): 56.

Provey, Joe. "The Sun Also Rises." *Popular Mechanics* (September 2002): 92.

Service, Robert F. "The Hydrogen Backlash." *Science* (August 13, 2004): 958-961.

"Stirrings in the Corn Fields." *The Economist* (May 12, 2005).

Terrell, Kenneth. "Running on Fumes." *U.S. News & World Report* (April 29, 2002): 58.

Tompkins, Joshua. "Dishing Out Real Power." *Popular Science* (February 1, 2005): 31.

Valenti, Michael. "Storing Hydroelectricity to Meet Peak-Hour Demand." *Mechanical Engineering* (April 1, 1992): 46.

Wald, Matthew L. "Questions about a Hydrogen Economy." *Scientific American* (May 2004): 66.

Westrup, Hugh. "Cool Fuel: Will Hydrogen Cure the Country's Addiction to Fossil Fuels?" *Current Science* (November 7, 2003): 10.

Westrup, Hugh. "What a Gas!" *Current Science* (April 6, 2001): 10.

WEB SITES

"Alternative Fuels." U.S. Department of Energy Alternative Fuels Data Center. http://www.eere.energy.gov/afdc/altfuel/altfuels.html (accessed on July 20, 2005).

"Alternative Fuels Data Center." U.S. Department of Energy: Energy Efficiency and Renewable Energy. http://www.eere.energy.gov/afdc/altfuel/p-series.html (accessed on July 11, 2005).

American Council for an Energy-Efficient Economy. http://aceee.org/ (accessed on July 27, 2005).

The American Solar Energy Society. http://www.ases.org/ (accessed on September 1, 2005).

Biodiesel Community. http://www.biodieselcommunity.org/ (accessed on July 27, 2005).

"Bioenergy." Natural Resources Canada. http://www.canren.gc.ca/tech_appl/index.asp?CaId=2&PgId=62 (accessed on July 29, 2005).

"Biofuels." Journey to Forever. http://journeytoforever.org/biofuel.html (accessed on July 13, 2005).

"Biogas Study." Schatz Energy Research Center. http://www.humboldt.edu/~serc/biogas.html (accessed on July 15, 2005).

"Black Lung." United Mine Workers of America. http://www.umwa.org/blacklung/blacklung.shtml (accessed on July 20, 2005).

"Classroom Energy!" American Petroleum Institute. http://www.classroom-energy.org (accessed on July 20, 2005).

"Clean Energy Basics: About Solar Energy." National Renewable Energy Laboratory. http://www.nrel.gov/clean_energy/solar.html (accessed on August 25, 2005).

"A Complete Guide to Composting." Compost Guide. http://www.compostguide.com/ (accessed on July 25, 2005).

"Conserval Engineering, Inc." American Institute of Architects. http://www.solarwall.com/ (accessed on September 1, 2005).

"The Discovery of Fission." Center for History of Physics. http://www.aip.org/history/mod/fission/fission1/01.html (accessed on December 17, 2005).

"Driving for the Future." California Fuel Cell Partnership. http://www.cafcp.org (accessed on August 8, 2005).

"Driving and Maintaining Your Vehicle." Natural Resources Canada. http://oee.nrcan.gc.ca/transportation/personal/driving/autosmart-maintenance.cfm?attr=11 (accessed on September 28, 2005).

"Ecological Footprint Quiz." Earth Day Network. http://www.earthday.net/footprint/index.asp (accessed on February 6, 2006).

"Energy Efficiency and Renewable Energy." U.S. Department of Energy. http://www.eere.energy.gov (accessed on September 28, 2005).

"Ethanol: Fuel for Clean Air." Minnesota Department of Agriculture. http://www.mda.state.mn.us/ethanol/ (accessed on July 14, 2005).

"Florida Solar Energy Center." University of Central Florida. http://www.fsec.ucf.edu (accessed on September 1, 2005).

"Fueleconomy.gov." United States Department of Energy. http://www.fueleconomy.gov/feg/ (accessed on July 27, 2005).

Fukada, Takahiro. "Japan Plans To Launch Solar Power Station In Space By 2040." *SpaceDaily.com,* Jan. 1, 2001. Available at http://www.spacedaily.com/news/ssp-01a.html (accessed Feb. 12, 2006).

"Geo-Heat Center." Oregon Institute of Technology. http://geoheat.oit.edu. (accessed on July 19, 2005).

"Geothermal Energy." World Bank. http://www.worldbank.org/html/fpd/energy/geothermal. (accessed on July 19, 2005).

Geothermal Resources Council. http://www.geothermal.org. (accessed on August 4, 2005).

"Geothermal Technologies Program." U.S. Department of Energy: Energy Efficiency and Renewable Energy. http://www.eere.energy.gov/geothermal. (accessed on July 22, 2005).

"Green Building Basics." California Home. http://www.ciwmb.ca.gov/GreenBuilding/Basics.htm (accessed on September 28, 2005).

"Guided Tour on Wind Energy." Danish Wind Industry Association. http://www.windpower.org/en/tour.htm (accessed on July 25, 2005).

"How the BMW H2R Works." How Stuff Works. http://auto.howstuffworks.com/bmw-h2r.htm (accessed on August 8, 2005).

"Hydrogen, Fuel Cells & Infrastructure Technologies Program." U.S. Department of Energy Energy Efficiency and Renewable Energy. http://www.eere.energy.gov/hydrogenandfuelcells/ (accessed on August 8, 2005).

"Hydrogen Internal Combustion." Ford Motor Company. http://www.ford.com/en/innovation/engineFuelTechnology/hydrogenInternalCombustion.htm (accessed on August 8, 2005).

"Incandescent, Fluorescent, Halogen, and Compact Fluorescent." California Energy Commission. http://www.consumerenergycenter.org/homeandwork/homes/inside/lighting/bulbs.html (accessed on September 28, 2005).

"Introduction to Green Building." Green Roundtable. http://www.greenroundtable.org/pdfs/Intro-To-Green-Building.pdf (accessed on September 28, 2005).

Lovins, Amory. "Mighty Mice: The most powerful force resisting new nuclear may be a legion of small, fast and simple micro-generation and efficiency projects." *Nuclear Engineering International,* Dec. 2005. Available at http://www.rmi.org/images/other/Energy/E05-15_MightyMice.pdf (accessed Feb. 12, 2006).

Nice, Karim. "How Hybrid Cars Work." Howstuffworks.com. http://auto.howstuffworks.com/hybrid-car.htm (accessed on September 28, 2005).

"Nuclear Terrorism—How to Prevent It." Nuclear Control Institute. http://www.nci.org/nuketerror.htm (accessed on December 17, 2005).

"Oil Spill Facts: Questions and Answers." *Exxon Valdez* Oil Spill Trustee Council. http://www.evostc.state.ak.us/facts/qanda.html (accessed on July 20, 2005).

O'Mara, Katrina, and Mark Rayner. "Tidal Power Systems." http://reslab.com.au/resfiles/tidal/text.html (accessed on September 13, 2005).

"Photos of El Paso Solar Pond." University of Texas at El Paso. http://www.solarpond.utep.edu/page1.htm (accessed on August 25, 2005).

"The Plain English Guide to the Clean Air Act." U.S. Environmental Protection Agency. http://www.epa.gov/air/oaqps/peg_caa/pegcaain.html (accessed on July 20, 2005).

"Reinventing the Automobile with Fuel Cell Technology." General Motors Company. http://www.gm.com/company/gmability/adv_tech/400_fcv/ (accessed on August 8, 2005).

"Safety of Nuclear Power." Uranium Information Centre, Ltd. http://www.uic.com.au/nip14.htm (accessed on December 17, 2005).

"Solar Energy for Your Home." Solar Energy Society of Canada Inc. http://www.solarenergysociety.ca/2003/home.asp (accessed on August 25, 2005).

The Solar Guide. http://www.thesolarguide.com (accessed on September 1, 2005).

"Solar Ponds for Trapping Solar Energy." United National Environmental Programme. http://edugreen.teri.res.in/explore/renew/pond.htm (accessed on August 25, 2005).

"Thermal Mass and R-value: Making Sense of a Confusing Issue." BuildingGreen.com. http://buildggreen.com/auth/article.cfm?fileName=070401a.xml (accessed on September 28, 2005).

"Tidal Power." University of Strathclyde. http://www.esru.strath.ac.uk/EandE/Web_sites/01-02/RE_info/Tidal%20Power.htm (accessed on September 13, 2005).

U.S. Department of Energy. *Report of the Review of Low Energy Nuclear Reactions*. Washington, DC: Department of Energy, Dec. 1, 2004. http://lenr-canr.org/acrobat/DOEreportofth.pdf, accessed Feb. 12, 2006.

Vega, L. A. "Ocean Thermal Energy Conversion (OTEC)." http://www.hawaii.gov/dbedt/ert/otec/index.html (accessed on September 13, 2005).

Venetoulis, Jason, Dahlia Chazan, and Christopher Gaudet. "Ecological Footprint of Nations: 2004." Redefining Progress. http://www.rprogress.org/newpubs/2004/footprintnations2004.pdf (accessed on February 8, 2006.)

Weiss, Peter. "Oceans of Electricity." *Science News Online* (April 14, 2001). http://www.science news.org/articles/20010414/bob12.asp (accessed on September 13, 2005).

"What Is Uranium? How Does It Work?" World Nuclear Association. http://www.world-nuclear.org/education/uran.htm (accessed on December 17, 2005).

"Wind Energy Tutorial." American Wind Energy Association. http://www.awea.org/faq/index.html (accessed on July 25, 2005).

Yam, Philip. "Exploiting Zero-Point Energy." *Scientific American*, December 1997. Available from http://www.padrak.com/ine/ZPESCIAM.html (accessed on August 2, 2005).

OTHER SOURCES

World Spaceflight News. *21st Century Complete Guide to Hydrogen Power Energy and Fuel Cell Cars: FreedomCAR Plans, Automotive Technology for Hydrogen Fuel Cells, Hydrogen Production, Storage, Safety Standards, Energy Department, DOD, and NASA Research*. Progressive Management, 2003.

Index

Italic type indicates volume number;* boldface type indicates entries and their pages; *(ill.) indicates illustrations.

B

C

E

G

I

J

K

L

M

N

T

U

V

W

Y

Z